Cambridge Tracts in Mathematics and Mathematical Physics

GENERAL EDITORS
F. SMITHIES, Ph.D. AND J. A. TODD, F.R.S.

No. 37

GEOMETRICAL OPTICS

T0269035

GEOMETRICAL OPTICS

AN INTRODUCTION
TO HAMILTON'S METHOD

by

J. L. SYNGE, Sc.D., F.R.S.

Senior Professor, School of Theoretical Physics,
Dublin Institute for Advanced Studies

CAMBRIDGE
AT THE UNIVERSITY PRESS
1962

CAMBRIDGE UNIVERSITY PRESS
Cambridge, New York, Melbourne, Madrid, Cape Town, Singapore, São Paulo, Delhi

Cambridge University Press
The Edinburgh Building, Cambridge CB2 8RU, UK

Published in the United States of America by Cambridge University Press, New York

www.cambridge.org
Information on this title: www.cambridge.org/9780521065900

First published 1937
Reprinted 1962
Re-issued in this digitally printed version 2008

A catalogue record for this publication is available from the British Library

ISBN 978-0-521-06590-0 paperback

CONTENTS

CHAPTER V
HETEROGENEOUS ISOTROPIC MEDIA

PREFACE

It is by no means easy for the applied mathematician to decide how much importance he should attach to the more abstract and aesthetic side of his work and how much to the detailed applications to physics, astronomy, engineering or the design of instruments. Great mathematical ideas do not blossom in workshops, as a rule, but on the other hand the theorist should not divorce himself from a healthy and intimate connection with practical questions.

Sir William Rowan Hamilton (1805–1865) created a method in Geometrical Optics, which, after lying long in disuse, is at last finding its proper place in the science. To all appearances, Hamilton attached little importance to the practical applications of his method, and it was only with the publication of his *Mathematical Papers*, Vol. I (Cambridge, 1931), that it was possible to form a more correct and balanced judgment of Hamilton as an applied mathematician. Great indeed was the labour which he employed with a view to applying his method to the design of optical instruments, but for him the abstract and aesthetic side of his work was of so much greater public importance than its practical use that the details of application remained unpublished till long after his death and long after other workers had discovered equivalent processes.

Since it was left largely to those primarily interested in optical design to develop the subject of Geometrical Optics, it is only natural that the student of the subject soon finds himself immersed in details which tend to cloud his understanding of the underlying general principles. Now, just as it is widely recognized that in the teaching of Mechanics a middle course must be steered between a completely abstract presentation and a technical approach, so it seems to me that the student of Geometrical Optics is most likely to understand the principles of Hamilton's method if he does not think too much at first of technical applications. But, at the same time, he should not be kept entirely remote from them.

Since editing, in collaboration with Professor A. W. Conway, F.R.S., Hamilton's papers on Geometrical Optics, I have had the opportunity of lecturing on the subject to graduate students and undergraduates in the University of Toronto. This book represents a course of twenty-five lectures to the latter. Although the reader may fail to find in it some things which he would naturally expect in a book on Geometrical Optics, no apology is offered on that account. If Hamilton's method is understood, the book serves its purpose. For that reason it is not necessary to defend the application of the method to problems which would admit shorter special solutions.

Hamilton was a master of mathematical notation, and he might in this respect be profitably studied by some modern writers in our subject. I have employed his notation in the main, changing the signs of the W and T functions to make their physical interpretation more obvious, and making some changes in nomenclature. It does not seem necessary or desirable to use the word "eikonal", which Bruns invented in 1895 in ignorance of Hamilton's work. Since one letter is just about as good as another, would it not be a harmless compliment to the genius of Hamilton for writers on Geometrical Optics to employ for the various characteristic functions the letters which he employed?

Although Hamilton himself started by considering the simpler case of isotropic media, it was not long before he saw that his method was also applicable to anisotropic media, and when he came to give his theory final form in his Third Supplement, he did so in all generality. This has done much to discourage those interested in the more practical aspects of his method, because in order to apply it they have been compelled to think in terms of (to them) unnecessary generality. To avoid a repetition of this error of policy, the theory of anisotropic media has been entirely omitted from this book. To compensate for this omission and for the fact that, although an attempt has been made to amplify

Hamilton's work in the directions since found of most interest, these amplifications have not been sufficient to create an adequate text-book, a brief bibliography is given below. In some of these works Hamilton's characteristic functions are referred to as Bruns' eikonals, but there is no significant difference. I have to thank three of my students, Messrs H. R. Roberts, P. R. Wallace and A. White, for assistance in the preparation of the manuscript, and my colleagues, Professor A. F. Stevenson and Dr B. A. Griffith, for reading the proofs and making valuable suggestions. It is also a pleasure to pay tribute to the skill and accuracy of the Cambridge University Press.

J. L. S.

TORONTO
October 1937

NOTE ON SECOND IMPRESSION

In this reprint only a few minor changes and corrections have been made.

J. L. S.

DUBLIN
July 1961

BIBLIOGRAPHY

R. S. Heath, *A Treatise on Geometrical Optics* (Cambridge University Press, 1895).

E. T. Whittaker, *The Theory of Optical Instruments* (Cambridge Tracts, No. 7, 1907).

J. G. Leathem, *The Elementary Theory of the Symmetrical Optical Instrument* (Cambridge Tracts, No. 8, 1908).

J. P. C. Southall, *The Principles and Methods of Geometrical Optics* (New York, 1913).

S. Czapski and O. Eppenstein, *Grundzüge der Theorie der optischen Instrumente* (Leipzig, 1924).

G. C. Steward, *The Symmetrical Optical System* (Cambridge Tracts, No. 25, 1928).

M. Herzberger, *Strahlenoptik* (Berlin, 1931).

The Mathematical Papers of Sir W. R. Hamilton, Vol. i (edited by A. W. Conway and J. L. Synge, Cambridge, 1931).

W. R. Hamiltons *Abhandlungen zur Strahlenoptik* (translated and edited with notes by G. Prange, Leipzig, 1933).

J. L. Synge, "Hamilton's Method in Geometrical Optics", *Journal of the Optical Society of America*, Vol. xxvii (1937), pp. 75-82.

M. Herzberger, *Modern Geometrical Optics* (Interscience, London, 1958).

M. Born and E. Wolf, *Principles of Optics* (Pergamon, London, 1959).

CHAPTER I

THE PRINCIPLES OF GEOMETRICAL OPTICS FOR ORDINARY MEDIA

1. The nature of geometrical optics.

A "perfect" scientific theory may be described as one which proceeds logically from a few simple hypotheses to conclusions which are in complete agreement with observation, to within the limits of accuracy of observation. But the theory is "useful" only in so far as it is possible to obtain conclusions from the hypotheses. As accuracy of observation increases, a theory ceases to be "perfect": modifications are introduced, making the theory more complicated and less "useful". Since we do not willingly surrender the wealth of approximate results furnished by the earlier form of the theory, we find ourselves in the unsatisfactory position of using one theory for one problem and another for another, although the two problems really belong to the same part of science. To rescue ourselves from intellectual confusion, we may admit theories called "ideal", in the sense that they deal with an ideal universe, resembling the actual universe to a fair degree of accuracy and usually corresponding to a limiting case of physical reality.

A critical examination of the history of mathematical physics shows that in truth man has always created "ideal" theories. Nature is much too complicated to be considered otherwise than in a simplified or idealized form, and it is inevitable that this idealization should lead to discrepancies between theoretical prediction and observation. As examples we may mention the mechanical theories of rigid bodies and perfect fluids; neither rigid bodies nor perfect fluids exist in nature. Or we may think of the Newtonian theory of gravitation, long regarded as "perfect", but now "ideal", physically replaced by the "perfect" (but not so "useful") general theory of relativity.

Geometrical optics is an ideal theory and a useful one. The discovery that the propagation of light is an electromagnetic phenomenon made the subject of optics coextensive with electromagnetism. We may, however, study certain parts of the subject of optics without reference to electromagnetism, always understanding that there is a limit to the physical accuracy of the results so obtained. It is customary to use the name "physical optics" for the more complex and physically accurate theory, and "geometrical optics" for the simpler ideal theory with which we shall be concerned. It is possible to justify geometrical optics as a limiting case of physical optics, the wave-length of the light in question tending to zero; † but we shall be content with the development of geometrical optics on the basis of its own hypotheses, just as it is customary to develop the dynamics of rigid bodies as a separate theory, and not as a limiting case of the dynamics of elastic bodies whose elastic moduli tend to infinity.

2. Fermat's principle: laws of reflection and refraction.

We consider the propagation of light through transparent media. We shall understand by an *ordinary* medium one which is *homogeneous* (the same at all points) and *isotropic* (the same for all directions) with respect to the propagation of light, deferring to Chapter v the discussion of media which are *heterogeneous* (like the atmosphere); *anisotropic* (crystalline) media will not be discussed.

Although the wave-length or frequency or colour of the light does not enter explicitly into the theory of geometrical optics, we admit that light is of various sorts. We shall consider separately the propagation of lights of different colours, so that at any one time we shall be dealing with *monochromatic* light.

In an ordinary medium light of a definite colour has a constant *velocity of propagation v*, different for different colours. In a vacuum, a particular case of an ordinary medium, the velocity of propagation is the same for all colours: it will be denoted by c. The *index of refraction* or *refractive index* of a medium is defined to be

$$(2\cdot1) \qquad \mu = c/v,$$

so that $\mu = 1$ for a vacuum. For all media $\mu \geqslant 1$. For most

† M. Born, *Optik* (Berlin, 1933), p. 45.

practical purposes air may be treated as a vacuum, because its index of refraction differs very little from unity: for air, the index (for the sodium D-line) is 1·0003.

If C is a curve passing through transparent media, joining points A' and A, the time which light would take to travel along C with velocity v would be

$$(2\cdot2) \qquad t = \int_{A'}^{A} \frac{ds}{v} = \frac{1}{c} \int_{A'}^{A} \mu\, ds,$$

where ds is an element of the curve. Since v is constant in each of the media (supposed ordinary), this may also be written

$$(2\cdot3) \qquad t = \sum_{A'}^{A} \frac{s}{v} = \frac{1}{c} \sum_{A'}^{A} \mu s,$$

where there is one term of the summation for each medium traversed, s being the length of C contained in it. We define the *optical length* of C to be

$$(2\cdot4) \qquad [C] = ct = \int_{A'}^{A} \mu\, ds = \sum_{A'}^{A} \mu s.$$

We shall use square brackets to indicate optical lengths.

We shall now state the basic hypothesis of geometrical optics:

Fermat's principle: When light travels from A' to A, it travels along a path or ray for which the time taken (or equivalently the optical length) has a stationary value with respect to infinitesimal variations of the path.

Usually the time will be a maximum or minimum.

Let us consider a single medium. Let A' and A be two points in it. Since the straight line joining A' and A has the shortest length of all possible curves joining these points, the optical length of this straight line (which only differs from the geometrical length by the constant factor μ) has a stationary value. Thus, *in a single ordinary medium light travels in straight lines.*

It would however be wrong to suppose that light can travel from A' to A only along the straight line $A'A$. It may pass from A' to the boundary of the medium and thence be reflected back to A. We shall now deduce the *law of reflection* from Fermat's principle.

Let light travel from A' to a surface S (Fig. 1) which bounds the medium in question, and hence back to A. S functions as a *mirror*. Let B be any point on S, so that $A'BA$ is in general an unnatural path (not a ray); its optical length is

(2·5) $$[A'BA] = \mu\rho' + \mu\rho,$$

where $\rho' = A'B$, $\rho = BA$. Let us take any rectangular axes of coordinates, and let the coordinates of the points be as follows:

(2·6) $\qquad A': x', y', z', \quad A: x, y, z, \quad B: x'', y'', z''.$

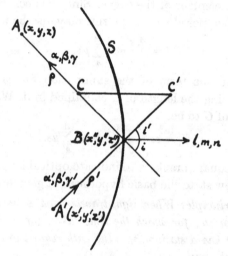

Fig. 1

Then

(2·7) $$\rho'^2 = \Sigma(x' - x'')^2, \quad \rho^2 = \Sigma(x - x'')^2,.$$

the Σ meaning a sum obtained by changing $x \to y \to z$. Giving to B an arbitrary infinitesimal displacement $\delta x''$, $\delta y''$, $\delta z''$, we have

(2·8) $\quad \rho'\,\delta\rho' = -\Sigma(x' - x'')\,\delta x'', \quad \rho\,\delta\rho = -\Sigma(x - x'')\,\delta x'',$

or if α', β', γ' are the direction cosines of $A'B$ and α, β, γ those of BA, so that

(2·9) $\quad \alpha'\rho' = x'' - x', \quad \beta'\rho' = y'' - y', \quad \gamma'\rho' = z'' - z',$

$\qquad\qquad \alpha\rho = x - x'', \quad \beta\rho = y - y'', \quad \gamma\rho = z - z'',$

we have

(2·10) $\qquad\qquad \delta\rho' = \Sigma\alpha'\,\delta x'', \quad \delta\rho = -\Sigma\alpha\,\delta x''.$

Thus for an arbitrary infinitesimal displacement of B on S, we have for the variation of the optical length

(2·11)
$$\delta[A'BA] = \mu\,\delta\rho' + \mu\,\delta\rho$$
$$= \mu\,\Sigma\,(\alpha' - \alpha)\,\delta x''.$$

In order that this may vanish, as demanded by Fermat's principle, for all arbitrary infinitesimal displacements of B on S, the vector whose components are

(2·12)
$$\alpha' - \alpha, \quad \beta' - \beta, \quad \gamma' - \gamma$$

must be parallel to the normal to S at B, or equivalently

(2·13)
$$\frac{\alpha' - \alpha}{l} = \frac{\beta' - \beta}{m} = \frac{\gamma' - \gamma}{n},$$

where l, m, n are the direction cosines of the normal to S at B: we shall take the normal drawn *into* the mirror as shown in Fig. 1.

The *angle of incidence* (i') is the angle between the incident ray produced and the normal to S, and the *angle of reflection* (i) is the angle between the reflected ray *reversed* and the normal. If we mark C' on the incident ray produced, and C on the reflected ray, making
$$BC' = BC = 1,$$

then the coordinates of C' relative to B are $(\alpha', \beta', \gamma')$ and those of C relative to B are (α, β, γ). Thus the vector (2·12) is the displacement CC': this is parallel to the normal at B. Hence it follows immediately that the *law of reflection* may be stated as follows:

   (i) the incident ray, the reflected ray and the normal to the mirror at the point of reflection are coplanar;

   (ii) the angle of incidence is equal to the angle of reflection $(i' = i)$.

It is easily seen that the common value of the fractions in (2·13) is $2\cos i$.

The analytic expression (2·13) for the law of reflection enables us to determine the reflected ray when the directions of the incident ray and the normal to the mirror at the point of incidence are known, for we have in (2·13) and the identity

$$(2·14) \qquad \alpha^2 + \beta^2 + \gamma^2 = 1,$$

three equations for the three unknowns α, β, γ. Since a quadratic equation occurs, there will be two solutions: the extraneous solution (to be rejected) is

$$\alpha = \alpha', \quad \beta = \beta', \quad \gamma = \gamma'.$$

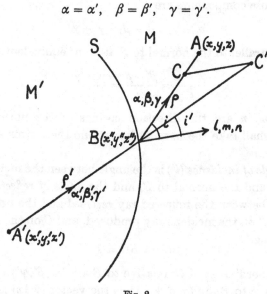

Fig. 2

When a ray passes from a medium M' of index μ' across a surface of separation S into a medium M of index μ, the ray in general undergoes an abrupt change of direction on crossing S, this phenomenon being known as *refraction*. Let us now investigate the law of refraction on the basis of Fermat's principle.

Let A' be a point in M', A a point in M and B any point on S (Fig. 2). Adopting the same notation as that used above for the case of reflection, we have for the optical length of the unnatural path $A'BA$

$$(2·15) \quad [A'BA] = \mu'\rho' + \mu\rho, \quad \rho' = A'B, \quad \rho = BA.$$

Then giving to B an arbitrary displacement $\delta x''$, $\delta y''$, $\delta z''$ on S, we have

(2·16)
$$\delta\,[A'BA] = \mu'\,\delta\rho' + \mu\,\delta\rho$$
$$= \mu'\,\Sigma\alpha'\,\delta x'' - \mu\,\Sigma\alpha\,\delta x''$$
$$= \Sigma\,(\mu'\alpha' - \mu\alpha)\,\delta x'',$$

and, since this is to vanish for the natural ray by Fermat's principle, we have (as the analogue to (2·13))

(2·17)
$$\frac{\mu'\alpha' - \mu\alpha}{l} = \frac{\mu'\beta' - \mu\beta}{m} = \frac{\mu'\gamma' - \mu\gamma}{n}.$$

If we lay off $BC' = \mu'$ along the incident ray produced, and $BC = \mu$ along the refracted ray, the coordinates of C' relative to B are $(\mu'\alpha', \mu'\beta', \mu'\gamma')$, and those of C relative to B are $(\mu\alpha, \mu\beta, \mu\gamma)$. Hence the numerators in (2·17) are the components of the displacement CC', which is therefore parallel to the normal to S at B.

The *angle of incidence* (i') is the angle between the incident ray produced and the normal to S (drawn from M' into M), and the *angle of refraction* (i) is the angle between the refracted ray and the normal to S. We may state the law of refraction as follows:

(i) the incident ray, the refracted ray and the normal to the refracting surface at the point of incidence are coplanar;

(ii) the angle of incidence and the angle of refraction are connected by the relation

(2·18) $\mu' \sin i' = \mu \sin i.$

This last relation follows at once by equating the lengths of the perpendiculars dropped from C' and C on the normal to S at B.

The common value of the fractions in (2·17) is

$$\mu' \cos i' - \mu \cos i.$$

As in the case of (2·13) for reflection, (2·17) (with (2·14)) give the direction of the refracted ray when the directions of the incident ray and the normal at the point of incidence are assigned. There is an extraneous solution arising from the quadratic equation involved. To see what it is, we choose special axes, the z-axis being normal to S at B, so that

$$(2\cdot19) \qquad\qquad l = m = 0, \quad n = 1.$$

Then $(2\cdot17)$ reduce to

$$(2\cdot20) \qquad\qquad \mu\alpha = \mu'\alpha', \quad \mu\beta = \mu'\beta'.$$

These determine α, β; γ is given by

$$(2\cdot21) \qquad\qquad \gamma = \pm\sqrt{1-\alpha^2-\beta^2}.$$

Obviously we must have $\gamma > 0$: the extraneous solution corresponds to the negative radical in $(2\cdot21)$, and hence the direction given by the extraneous solution is the geometrical reflection in the tangent plane to S at B of the true refracted ray.

Under certain circumstances it is impossible for a ray from M' to be refracted into M. Then reflection only, and not refraction, can take place: this phenomenon is known as *total reflection*. It occurs when i cannot be found to satisfy $(2\cdot18)$, that is, when

$$(2\cdot22) \qquad\qquad \frac{\mu'}{\mu}\sin i' > 1.$$

Obviously total reflection can take place only if $\mu' > \mu$.

3. Normal and skew congruences: theorem of Malus.

A system of curves filling a portion of space, and such that in general a single curve passes through any assigned point, is called a *congruence*. For example, the normals to a surface form a congruence. If we denote by α, β, γ the direction cosines of the tangent to the curve of the congruence at a point x, y, z, the congruence may be defined by expressing α, β, γ as functions of x, y, z,

$$(3\cdot1) \quad \alpha = f(x,y,z), \quad \beta = g(x,y,z), \quad \gamma = h(x,y,z).$$

These three functions are not independent; they must satisfy the identity

$$(3\cdot2) \qquad\qquad f^2 + g^2 + h^2 = \alpha^2 + \beta^2 + \gamma^2 = 1.$$

If the curves which form the congruence are straight lines, the congruence is said to be a *rectilinear congruence*. The congruences with which we have to deal in the geometrical optics of homogeneous media are rectilinear.

If there exists a singly infinite family of surfaces cut orthogonally by the curves of a congruence, the congruence is said to be *normal*; if such a family does not exist, the congruence is said to be *skew*.

Suppose there is a normal congruence of curves, defined as in (3·1). Let the equations of the normal surfaces be expressed in the form

(3·3) $F(x, y, z) = \text{const.}$

The direction cosines of the normal to the surface of this family at a point x, y, z have the ratios

$$\frac{\partial F}{\partial x} : \frac{\partial F}{\partial y} : \frac{\partial F}{\partial z}.$$

Therefore

(3·4) $\theta\alpha = \dfrac{\partial F}{\partial x}, \quad \theta\beta = \dfrac{\partial F}{\partial y}. \quad \theta\gamma = \dfrac{\partial F}{\partial z},$

where θ is a factor of proportionality. Differentiating, we have

$$\frac{\partial}{\partial y}(\theta\gamma) = \frac{\partial^2 F}{\partial y\,\partial z} = \frac{\partial^2 F}{\partial z\,\partial y} = \frac{\partial}{\partial z}(\theta\beta),$$

and therefore

(3·5) $\theta\left(\dfrac{\partial\gamma}{\partial y} - \dfrac{\partial\beta}{\partial z}\right) + \gamma\dfrac{\partial\theta}{\partial y} - \beta\dfrac{\partial\theta}{\partial z} = 0.$

Similarly, we obtain

(3·6)
$$\begin{cases} \theta\left(\dfrac{\partial\alpha}{\partial z} - \dfrac{\partial\gamma}{\partial x}\right) + \alpha\dfrac{\partial\theta}{\partial z} - \gamma\dfrac{\partial\theta}{\partial x} = 0, \\[2ex] \theta\left(\dfrac{\partial\beta}{\partial x} - \dfrac{\partial\alpha}{\partial y}\right) + \beta\dfrac{\partial\theta}{\partial x} - \alpha\dfrac{\partial\theta}{\partial y} = 0. \end{cases}$$

Multiplying these equations in order by α, β, γ, adding, and dividing by θ, we obtain

(3·7) $\alpha\left(\dfrac{\partial\gamma}{\partial y} - \dfrac{\partial\beta}{\partial z}\right) + \beta\left(\dfrac{\partial\alpha}{\partial z} - \dfrac{\partial\gamma}{\partial x}\right) + \gamma\left(\dfrac{\partial\beta}{\partial x} - \dfrac{\partial\alpha}{\partial y}\right) = 0.$

This condition is necessarily satisfied if the congruence is normal. Moreover it is known from the theory of total differential equations† that if (3·7) is satisfied, then functions θ and F of

† Cf. H. T. H. Piaggio, *Differential Equations* (London, 1933), p. 140.

x, y, z exist such that (3·4) are true. In other words, the equation

$$\alpha\, dx + \beta\, dy + \gamma\, dz = 0$$

is integrable. Consequently (3·7), if true, implies the existence of a family of surfaces (3·3) to which the curves of the congruence are normal. *Therefore* (3·7) *is a necessary and sufficient condition that a congruence be normal.*

As an example, it is easily verified that the congruence defined by

$$\alpha = x/r, \quad \beta = y/r, \quad \gamma = z/r, \quad (r^2 = x^2 + y^2 + z^2),$$

is a normal congruence. On the other hand, it may be shown that the congruence defined by

$$\alpha = y/r, \quad \beta = -x/r, \quad \gamma = z/r$$

is a skew congruence.

If we are given a family of surfaces

$$F(x, y, z) = \text{const.},$$

there exists a normal congruence of which these surfaces are the normal surfaces. The curves of this congruence are called the *orthogonal trajectories* of the family of surfaces. The congruence has the equations

(3·8) $$\alpha = \phi\,\frac{\partial F}{\partial x}, \quad \beta = \phi\,\frac{\partial F}{\partial y}, \quad \gamma = \phi\,\frac{\partial F}{\partial z},$$

where

(3·9) $$\phi = \frac{1}{\sqrt{\left(\dfrac{\partial F}{\partial x}\right)^2 + \left(\dfrac{\partial F}{\partial y}\right)^2 + \left(\dfrac{\partial F}{\partial z}\right)^2}}.$$

We shall now show that *the system of straight lines normal to any assigned surface is a normal congruence.* Let S_0 (Fig. 3) be the given surface and let S be the surface formed by cutting off the same length s from all the normals to S_0. This construction puts the points of S_0 and S into one-to-one correspondence. Let A, B be the points of S corresponding to A_0, B_0 on S_0, respectively, the distance $A_0 B_0$ being infinitesimal. Join A_0 to B. Since BB_0 is normal to S_0 at B_0, the infinitesimal displace-

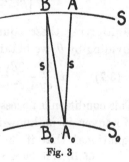

Fig. 3

ment B_0A_0 is perpendicular to B_0B. Hence

$$A_0B = B_0B = s = A_0A,$$

to the first order of infinitesimals. Thus the infinitesimal triangle BA_0A is isosceles to the first order, and therefore the angle BAA_0 is a right angle: thus A_0A is normal to S at A. Thus all the normals to S_0 are normals to S, and this is true for any value of s. Hence the congruence of normals to S_0 is a normal congruence.†

It is obvious that the rays emanating from a point P in an ordinary medium form a normal rectilinear congruence, having for normal surfaces the family of spheres with centre P. The theorem of Malus asserts that *a normal rectilinear congruence remains normal after reflection or refraction*, and hence that the congruence formed by any number of reflections or refractions from the congruence of rays originally emanating from a point is a normal congruence.

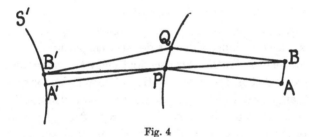

Fig. 4

To prove the theorem of Malus, let $A'P$ be a ray of the incident congruence, incident on the reflecting or refracting surface at P, and let PA be the reflected or refracted ray (Fig. 4). Let S' be the normal surface to the incident congruence (normal by hypothesis) at A'; let B' be an adjacent point on S' and $B'QB$ the ray through B', incident at Q. The point B is taken so that

(3·10) $$[B'QB] = [A'PA].$$

Joining $B'P, PB$, we have to the first order, by Fermat's principle,

(3·11) $$[B'PB] = [B'QB].$$

† The result also follows immediately from the equation

$$\Sigma (x - x_0)(\delta x - \delta x_0) = 0,$$

where x, y, z are coordinates of A and x_0, y_0, z_0 those of A_0.

Hence

(3·12) $[B'P]+[PB] = [A'P]+[PA];$

but since $A'B'$ is perpendicular to $A'P$, we have $[B'P] = [A'P]$, to the first order, and hence $[PB] = [PA]$. Thus to the first order, $PB = PA$, which shows that the infinitesimal displacement AB is perpendicular to PA.

Now if a surface S is formed by taking points on the reflected or refracted rays so that the optical length from each point on S' to its correspondent on S is $[A'PA]$, it follows from the result established above that PA is perpendicular to every infinitesimal displacement at A on S. Therefore PA is normal to S at A. Thus all the reflected or refracted rays are normal to S. By varying the position of A on the reflected or refracted ray from A', we get a single infinity of surfaces to which the final rays are normal. Thus the theorem of Malus is established.

4. The construction of Huyghens.

Throughout the history of the science of optics, two rival theories have developed side by side—the corpuscular theory and the wave theory. In the corpuscular theory the phenomenon of light is regarded as due to the motion of corpuscles (or quanta in modern language), which are individually localized in small regions of space, so that ideally they may be regarded as points. The tracks of these particles are the rays. In the wave theory, light is regarded as due to the propagation of a system of waves. At the present time it is impossible to be dogmatic concerning the correctness of either view. However, in geometrical optics it is possible to regard the two theories as different aspects of a single theory.

To develop the wave theory of light in geometrical optics we follow the construction of Huyghens. Let Σ' be a surface which represents a wave of disturbance in the medium at time t'. Let each point A' of Σ' be regarded as the centre of a secondary disturbance which spreads out from A' in all directions. For an anisotropic medium it is necessary to distinguish between the *ray-velocity* and the *wave-velocity*, but we are here concerned only with ordinary media, in which it is assumed that the velocity of

propagation of the secondary waves is the same as the ray-velocity
v introduced in § 2.

Let us now consider the disturbance at time t: we have then a
number of secondary waves, each of radius $v(t-t')$, whose points
fill a layer of space including the surface Σ'. These spheres have
an envelope consisting of two sheets, one sheet on each side of Σ'.
We assume that the wave Σ' has a *sense of propagation* to one side
or the other, and we assume that *the wave at time t is that sheet Σ*
of the envelope of the secondary waves which is such that passage
from Σ' to Σ is in the assumed sense of propagation. In Fig. 5,

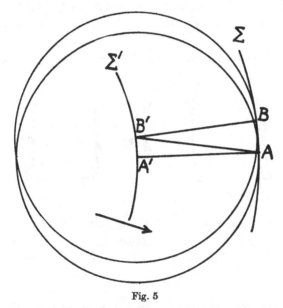

Fig. 5

A', B' are centres of secondary waves and A, B the points of
contact of these waves with the envelope Σ.

In the above statement it is assumed that the secondary waves
do not cut the boundary of the medium. To deal with such cases,
in which reflection or refraction takes place, we have to proceed
by infinitesimal steps. When a point of the wave lies on a boundary
of the medium, it becomes the centre of a secondary wave which,
in the case of reflection, has a sense of propagation back into the

medium, and, in the case of refraction, goes on into the second medium, but with a different velocity. It is evident that the construction of Huyghens leads to definite reflected and refracted waves. These will be discussed below.

First, however, let us consider the propagation of a wave in a single medium, without reflection. We shall establish the following facts concerning propagation according to the construction of Huyghens. *Given a wave Σ' at time t', the resultant wave Σ at time t is the same whether developed by the construction of Huyghens in one step or in several steps: also, the normals to Σ' are normal to Σ, and if $A'A$ is one of these normals, with A' on Σ' and A on Σ, then A is the point of contact with Σ of the secondary wave having its centre at A'.*

Fig. 6 shows the wave Σ reached in four steps from Σ'; $\Sigma_1, \Sigma_2, \Sigma_3$

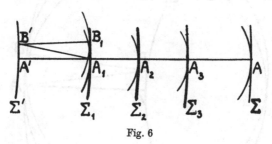

Fig. 6

are the intermediate waves. A' is the centre of a secondary wave which touches Σ_1 at A_1. B' is adjacent to A' on Σ' and its secondary wave touches Σ_1 at B_1. It is implied in the construction that

(4·1) $A'A_1 = B'B_1.$

Since the spheres touch their envelope, we have, to the first order,

(4·2) $B'B_1 = B'A_1;$

hence by (4·1)

(4·3) $A'A_1 = B'A_1,$

which shows that $A'A_1$ is normal to Σ'. It is of course also normal to Σ_1 since, as radius of the sphere with centre A', it is normal to the tangent plane to that sphere at A_1, and this plane is also the tangent plane of Σ_1. Hence $A'A_1$ is normal to both the waves which it connects, and its length is $v(t_1 - t')$, if t', t_1 are the times for Σ', Σ_1 respectively. Continuing the construction, we see that $A_1 A_2$, being normal to Σ_1, lies in the same straight line with $A'A_1$; it is of length $v(t_2 - t_1)$, where t_2 is the time for Σ_2. Hence we see that, by the application of the four steps shown, the secondary waves with centres A', A_1, A_2, A_3 lead finally to a contact A with the final surface Σ such that A lies on the normal to Σ' at A' at a distance

$$(4\cdot4) \quad A'A = v(t_1 - t') + v(t_2 - t_1) + v(t_3 - t_2) + v(t - t_3) = v(t - t'),$$

where t is the time for Σ. Since $A_3 A$ is normal to Σ, so also is $A'A$. But it is now evident that if only one step for the time

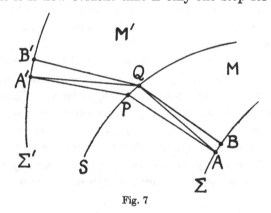

Fig. 7

interval $t - t'$ were employed, we would get a wave such that its point of contact with the secondary wave with centre A' would lie on the normal to Σ' at A' at a distance $v(t - t')$, i.e. precisely at the point A. The result italicized above is therefore established.

Let us now consider reflection or refraction according to the construction of Huyghens. Since the treatments for reflection and refraction are almost identical, it will suffice to consider refraction.

Light travels from a medium M' into a medium M across a surface S (Fig. 7). Σ' and Σ represent positions of a wave at times t' and t respectively; P and Q are any two adjacent points on

S; $A'P$, $B'Q$ are normals to the family of incident waves, and PA, QB normals to the family of refracted waves, A', B' lying on Σ' and A, B on Σ. Now we know from the construction of Huyghens that the times taken to traverse $B'QB$ and $A'PA$ are the same, or in terms of optical lengths

$$(4\cdot5) \qquad [B'QB] = [A'PA].$$

But from the normal property, we have to the first order

$$(4\cdot6) \qquad [B'Q] = [A'Q], \quad [QB] = [QA],$$

and hence

$$(4\cdot7) \qquad [A'QA] = [A'PA]:$$

in fact the optical length measured along the normals to the waves from A' to A has a stationary value. Hence (2·17) may be established as the law of refraction for wave-normals by applying the stationary condition as in § 2.

Seeing that reflection may be treated in the same way, we may state the following result, which reconciles the ray theory and the wave theory in geometrical optics, as far as ordinary media are concerned. *Given a wave Σ' in a medium M', there is determined by the construction of Huyghens a system of waves Σ after reflection or refraction. Given a system of rays normal to Σ' in M', there is determined by Fermat's principle a system of rays after reflection or refraction. This latter system of rays is normal to the waves Σ.*

If light starts from a point source A', the waves are spheres having A' for centre; the rays are the radii, normal to the spheres. By the result just established, the normality of rays and waves is conserved over each reflection and refraction. Thus if we think simultaneously of the rays and the surfaces to which they are normal, we have in mind at the same time the two theories of rays and waves.

CHAPTER II

THE CHARACTERISTIC FUNCTIONS FOR INSTRUMENTS FORMED OF ORDINARY MEDIA

5. The characteristic function V.

Let us consider an instrument formed of $n+1$ media with indices of refraction
$$\mu', \mu_1, \mu_2, \ldots, \mu_{n-1}, \mu,$$
separated by surfaces $\quad S_1, S_2, \ldots, S_n$.

It is simplest to suppose that only refractions take place: if a reflection takes place, the medium in which it occurs is counted twice over, the same analysis applying.

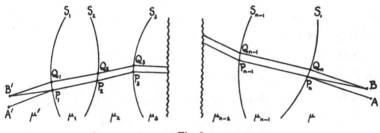

Fig. 8

Let $A'P_1 \ldots P_n A$ be a ray traversing the instrument, A' lying in the first medium and A in the last. By Fermat's principle we know that this ray has a stationary optical length when compared with adjacent unnatural paths joining A' to A.

Let $Oxyz$ be rectangular axes of coordinates.† Let x', y', z' be the coordinates of A' and x, y, z those of A. If these six numbers are given, points A', A are determined; hence by Fermat's principle a ray $A'A$ and a corresponding optical length are also determined.‡ *The characteristic function* $V(x', y', z', x, y, z)$ *of the*

† We might employ different axes for the initial and final media, but since it is at times necessary to use a single set of coordinates for both media, we shall, to avoid confusion, use a single set throughout.

‡ It may happen that there is no ray joining A' and A: then V is not defined for that pair of points.

instrument is defined to be the optical length of the ray from the point $A'(x', y', z')$ to the point $A(x, y, z)$:

(5·1) $\qquad V(x', y', z', x, y, z) = [A'P_1 \dots P_n A]$.

To distinguish it from the other characteristic functions to be defined later, V may be called the *point-characteristic.*

Passing from the ray $A'A$ to an adjacent ray $B'Q_1 \dots Q_n B$, and denoting the coordinates as follows:

$$B': \quad x'+\delta x', \quad y'+\delta y', \quad z'+\delta z',$$
$$B: \quad x+\delta x, \quad y+\delta y, \quad z+\delta z,$$

we have, for the increment in V,

(5·2) $\quad \delta V = V_{x'}\delta x' + V_{y'}\delta y' + V_{z'}\delta z' + V_x \delta x + V_y \delta y + V_z \delta z$
$$\qquad\qquad = \Sigma V_{x'} \delta x' + \Sigma V_x \delta x,$$

the subscripts denoting partial derivatives. But by Fermat's principle we have

(5·3) $\qquad [B'P_1 \dots P_n B] = [B'Q_1 \dots Q_n B]$,

to the first order, and hence

(5·4) $\qquad \delta V = [B'Q_1 \dots Q_n B] - [A'P_1 \dots P_n A]$
$$\qquad\qquad = [B'P_1 \dots P_n B] - [A'P_1 \dots P_n A]$$
$$\qquad\qquad = [B'P_1] - [A'P_1] + [P_n B] - [P_n A]$$
$$\qquad\qquad = \mu' \, \delta\rho' + \mu \, \delta\rho,$$

where

(5·5) $\quad \rho' = A'P_1,\ \rho'+\delta\rho' = B'P_1;\quad \rho = P_n A,\ \rho+\delta\rho = P_n B$.

Let

(5·6) $\qquad \begin{cases} \alpha', \beta', \gamma' = \text{direction cosines of } A'P_1, \\ \alpha, \ \beta, \ \gamma \ = \text{direction cosines of } P_n A. \end{cases}$

Then, as in § 2, we see that

(5·7) $\qquad \begin{cases} \delta\rho' = -(\alpha'\,\delta x' + \beta'\,\delta y' + \gamma'\,\delta z') = -\Sigma\alpha'\,\delta x', \\ \delta\rho \ = \alpha\,\delta x + \beta\,\delta y + \gamma\,\delta z = \Sigma\alpha\,\delta x, \end{cases}$

and hence, by (5·4),

(5·8) $\qquad\qquad \delta V = -\mu'\,\Sigma\alpha'\,\delta x' + \mu\,\Sigma\alpha\,\delta x$.

Comparing (5·2) and (5·8), and noting that therefore

(5·9) $\qquad \Sigma V_{x'}\,\delta x' + \Sigma V_x\,\delta x = -\mu'\,\Sigma\alpha'\,\delta x' + \mu\,\Sigma\alpha\,\delta x$

for arbitrary values of the infinitesimals,† we see that

(5·10)
$$\begin{cases} V_{x'} = -\mu'\alpha', & V_{y'} = -\mu'\beta', & V_{z'} = -\mu'\gamma', \\ V_x = \mu\alpha, & V_y = \mu\beta, & V_z = \mu\gamma. \end{cases}$$

It is convenient to introduce the *components* of the initial and final rays, defined by

(5·11)
$$\begin{cases} \sigma' = \mu'\alpha', & \tau' = \mu'\beta', & \upsilon' = \mu'\gamma', \\ \sigma = \mu\alpha, & \tau = \mu\beta, & \upsilon = \mu\gamma. \end{cases}$$

In terms of them, (5·10) may be written

(5·12)
$$\begin{cases} V_{x'} = -\sigma', & V_{y'} = -\tau', & V_{z'} = -\upsilon', \\ V_x = \sigma, & V_y = \tau, & V_z = \upsilon. \end{cases}$$

We note that between the components there exist the identical relations

(5·13) $\qquad \sigma'^2 + \tau'^2 + \upsilon'^2 = \mu'^2, \qquad \sigma^2 + \tau^2 + \upsilon^2 = \mu^2.$

Hence, by (5·10) or (5·12), it follows that the characteristic function V satisfies the two partial differential equations‡

(5·14) $\qquad V_{x'}^2 + V_{y'}^2 + V_{z'}^2 = \mu'^2, \qquad V_x^2 + V_y^2 + V_z^2 = \mu^2.$

In dealing with the behaviour of an optical instrument we have under consideration primarily the following twelve quantities:

(5·15)
$$\begin{cases} x', y', z', \text{ coordinates of a point on the initial ray,} \\ \sigma', \tau', \upsilon', \text{ components of the initial ray,} \\ x, y, z, \quad \text{coordinates of a point on the final ray,} \\ \sigma, \tau, \upsilon, \quad \text{components of the final ray.} \end{cases}$$

These twelve quantities are not all independent on account of the two identities (5·13).

The following questions may be asked:

(a) Given the coordinates of initial and final points, what are the components at them of the ray passing through these points?·

† In certain special cases these six infinitesimals are not independent: this will occur when the congruence of rays from a point B', chosen arbitrarily in the neighbourhood of A', fails to pass through all points of a three-dimensional region containing A. We shall, for simplicity, exclude such cases from consideration.

‡ Hamilton's dynamical theory was very closely related to his optical theory. Either of the equations (5·14) will be recognized as the Hamilton-Jacobi equation for a particle moving under no forces.

(b) Given an initial point and the components of an initial ray through it, what are the components of the final ray and the coordinates of a point on it?

(c) The same as (b), with interchange of the words "initial" and "final".

(d) Given an initial point and the components of a final ray, what are the coordinates of a final point and the components of the initial ray?

(e) The same as (d), with interchange of the words "initial" and "final".

(f) Given the components of initial and final rays, what are the coordinates of points on them? Or, in other words, where are the rays situated?

Later methods will show us how to answer (d), (e), (f). For the present we remark that if the characteristic function V of an instrument is known, the equations (5·12) immediately supply the answer to question (a). The difficulty in the useful application of (5·12) lies in the difficulty of calculating V for an actual instrument.

The function V defined above is a function of six variables, namely, the coordinates of initial and final points. It is defined by the *instrument*. If we are merely interested in the final congruence of rays due to a source at a fixed point x', y', z', it is no longer necessary to emphasize the dependence of V on x', y', z', and we may consider it as a function of x, y, z. We then think of $V(x, y, z)$ as the *characteristic function of the final congruence of rays*, the components of these rays being, as in (5·12),

(5·16) $\sigma = V_x, \quad \tau = V_y, \quad \upsilon = V_z.$

The characteristic function for a normal congruence of rays in a medium of index μ may also be defined in a slightly different, but essentially equivalent, manner as follows. Let Σ (Fig. 9) be any normal surface of the congruence and let $P(x, y, z)$ be any point. Let MP be the ray through P, M being on Σ. Let us define

(5·17) $V^* = [MP] = \mu MP;$

V^* is a function of x, y, z. If we displace P to
$$Q(x + \delta x, y + \delta y, z + \delta z),$$
we have

(5·18)
$$\delta V^* = \Sigma V_x^* \, \delta x = [NQ] - [MP] = \mu(NQ - MP) = \mu \, \delta MP.$$

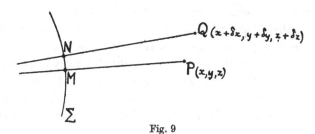

Fig. 9

Since the projection of NQ on MP is (to the first order) equal to NQ and since MN is perpendicular to MP, δMP is equal to the projection of PQ on MP, and so

(5·19) $$\delta MP = \Sigma \alpha \, \delta x,$$

where α, β, γ are the direction cosines of MP. Hence, the components being defined as before by (5·11), we have

(5·20) $$\delta V^* = \mu \Sigma \alpha \, \delta x = \Sigma \sigma \, \delta x,$$

which leads us at once to (5·16), with V replaced by V^*. It is easily seen that when the final congruence of rays comes originally from a point source, the V^* as just defined differs only by a constant from the V discussed earlier.

It is possible to design a mirror to reflect to an assigned point A all rays of a normal congruence. For if Σ (Fig. 9) is a normal surface to the congruence, such a mirror is given by the locus of a point P such that $$MP + PA = \text{const.}$$

The proof follows at once from Fermat's principle. Similarly, we can design a refracting surface of material of any assigned index to bring a given normal congruence after refraction to an assigned point A in the material. These mirrors and refracting surfaces may be called *focal reflectors* and *refractors*.

Similarly, a mirror or refracting surface may be found to turn any normal congruence into a parallel congruence in an arbitrarily assigned direction.

Let us return to the general point of view, according to which V is regarded as a function of six variables. According to the definition it would appear that x', y', z' may be the coordinates of any point in the initial medium and x, y, z the coordinates of any point in the final medium. Actually, however, these points cannot range right through their respective media, because it is implied in the definition that it is possible for a ray to pass from the one to the other, and this will not in general be the case for *all* pairs of points. Thus in general the ranges of x', y', z' and of x, y, z are only parts of the initial and final media respectively.

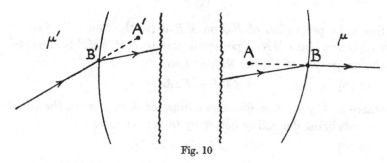

Fig. 10

It is, however, possible and convenient to "continue" the function V for values of x', y', z', x, y, z corresponding to points which do not lie in the initial or final media, but lie on initial or final rays produced. Thus, in Fig. 10, $A'(x',y',z')$ lies on the initial ray produced and $A(x,y,z)$ on the final ray produced backwards. We may proceed from A' to A by first going along $A'B'$, the production of the initial ray, then through the instrument from B' to B, and then along BA, the production of the final ray. We define the optical length of this route from A' to A as

(5·21) $$[A'B'] + [B'B] + [BA],$$

where $[A'B']$ is an optical length, calculated as if the index were μ', that of the initial medium, and counted negative because described in a negative sense, and $[BA]$ similarly calculated as if the index were μ, that of the final medium, and also counted negative. We then define $V(x', y', z', x, y, z)$ as the optical length (5·21) computed in this way. It is easily seen that Fermat's principle holds for optical lengths interpreted as above, and furthermore that the fundamental relations (5·12) also hold for the function V continued in this way.

As remarked above, the utility of the function V is restricted, owing to the difficulty of calculating it. For such a simple instrument as a plane mirror, however, we can write it down from elementary considerations. If the mirror is $z = 0$, we have

$$(5\cdot22) \qquad V = \sqrt{(x'-x)^2 + (y'-y)^2 + (z'+z)^2}.$$

For a set of three mirrors at right angles to one another, coincident with the planes $x = 0$, $y = 0$, $z = 0$, we have

$$(5\cdot23) \qquad V = \sqrt{(x'+x)^2 + (y'+y)^2 + (z'+z)^2}.$$

These simple results may be deduced by the elementary method of images. To calculate V for a general instrument, as in Fig. 8, we proceed as follows. Let x_i, y_i, z_i be the running coordinates of a point on the surface S_i, and let the equation of this surface be

$$(5\cdot24) \qquad f_i(x_i, y_i, z_i) = 0.$$

Let us draw any path of straight segments from $A'(x', y', z')$ to $A(x, y, z)$: if $P_1, P_2, ..., P_n$ are the points where this path meets the surfaces, its optical length is

$$(5\cdot25) \qquad L = \mu' A' P_1 + \sum_{i=1}^{n-1} \mu_i P_i P_{i+1} + \mu P_n A.$$

This can be expressed easily as a function of

$$(5\cdot26) \qquad x', y', z', x, y, z, x_i, y_i, z_i \qquad (i = 1, 2, ..., n).$$

By Fermat's principle we know that L has a stationary value for the natural ray for small variations of $P_1, ..., P_n$ on their respective surfaces. Thus

$$(5\cdot27) \qquad \sum_{i=1}^{n} \left(\frac{\partial L}{\partial x_i} \delta x_i + \frac{\partial L}{\partial y_i} \delta y_i + \frac{\partial L}{\partial z_i} \delta z_i \right) = 0$$

if

(5·28) $\quad \dfrac{\partial f_i}{\partial x_i}\,\delta x_i + \dfrac{\partial f_i}{\partial y_i}\,\delta y_i + \dfrac{\partial f_i}{\partial z_i}\,\delta z_i = 0 \quad (i = 1, 2, ..., n).$

Consequently

(5·29) $\quad \dfrac{\partial L}{\partial x_i} = \lambda_i \dfrac{\partial f_i}{\partial x_i}, \quad \dfrac{\partial L}{\partial y_i} = \lambda_i \dfrac{\partial f_i}{\partial y_i}, \quad \dfrac{\partial L}{\partial z_i} = \lambda_i \dfrac{\partial f_i}{\partial z_i}$

$$(i = 1, 2, ..., n),$$

the λ's being undetermined multipliers. In (5·24) and (5·29) we have $4n$ equations for the $4n$ quantities $x_i, y_i, z_i, \lambda_i\ (i = 1, 2, ..., n)$, and if these quantities are found, and the values of x_i, y_i, z_i substituted in (5·25), we have the characteristic function

(5·30) $\qquad\qquad V(x', y', z', x, y, z) = L.$

Although theoretically simple, the solutions or eliminations demanded by the method usually prove very difficult; the functions W and T to be defined later are easier to calculate as a rule.

6. The characteristic function W.

Consider a ray passing through an instrument. Let A' be a point on the initial ray and N the foot of the perpendicular dropped from the origin O on the final ray (Fig. 11)†. We define W by

(6·1) $\qquad\qquad W = [A'N].$

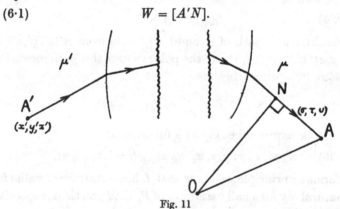

Fig. 11

If A' is assigned as a source of light, there will in general be final

† Different axes may be used for the initial and final media, but for simplicity we shall employ a single set of axes for both media.

rays with all directions in a certain range of directions, and at most a finite number of final rays with an assigned direction. Thus if x', y', z' are the coordinates of A' and σ, τ, v the components of the final ray (connected of course by the identity

$$(6 \cdot 2) \qquad \sigma^2 + \tau^2 + v^2 = \mu^2,$$

so that v is determined by σ, τ to within an ambiguous sign), we may say that $\dot{W}$ is a function (possibly multiple-valued) of the variables x', y', z', σ, τ, or

$$(6 \cdot 3) \qquad W = W(x', y', z', \sigma, \tau).$$

W is to be regarded as a second *characteristic function* of the instrument. It may be called the *mixed characteristic*.

It is evident that the continuation process described in § 5 enables us to take initial points x', y', z' which do not lie in the initial medium, but on the initial rays produced, and to employ the above definition for W even though the perpendicular ON falls not on the final ray but on the final ray produced backwards.

We shall now show the connection between V and W. Let $A(x, y, z)$ be any point on the final ray. Then it is easily seen by orthogonal projection of OA on the final ray that

$$(6 \cdot 4) \qquad [NA] = \mu \Sigma \alpha x = \Sigma \sigma x.$$

Hence

$$(6 \cdot 5) \qquad V(x', y', z', x, y, z) = [A'A] = [A'N] + [NA]$$
$$= W(x', y', z', \sigma, \tau) + (\sigma x + \tau y + vz).$$

Let us now give arbitrary variations to A' and A: this causes variations in the components of the final ray. Differentiating $(6 \cdot 5)$, we have

$$(6 \cdot 6) \quad \Sigma V_{x'} \delta x' + \Sigma V_x \delta x$$
$$= \Sigma W_{x'} \delta x' + W_\sigma \delta\sigma + W_\tau \delta\tau + \Sigma\sigma \delta x + \Sigma x \delta\sigma.$$

By $(5 \cdot 12)$ we have

$$(6 \cdot 7) \qquad \Sigma V_{x'} \delta x' = -\Sigma\sigma' \delta x', \quad \Sigma V_x \delta x = \Sigma\sigma \delta x,$$

and by $(6 \cdot 2)$

$$(6 \cdot 8) \qquad \delta v = -\frac{\sigma \delta\sigma + \tau \delta\tau}{v}.$$

Thus (6·6) becomes

(6·9) $-\Sigma\sigma'\,\delta x' = \Sigma W_{x'}\,\delta x' + W_\sigma\,\delta\sigma + W_\tau\,\delta\tau$
    $+ (x - z\sigma/v)\,\delta\sigma + (y - z\tau/v)\,\delta\tau.$

But $\delta x'$, $\delta y'$, $\delta z'$, $\delta\sigma$, $\delta\tau$ are arbitrary and independent.† Hence

(6·10) $\sigma' = -W_{x'}, \quad \tau' = -W_{y'}, \quad v' = -W_{z'},$

(6·11) $x - z\sigma/v = -W_\sigma, \quad y - z\tau/v = -W_\tau.$

We are now in a position to answer question (d) raised in § 5, if we suppose the function W known.‡ For given an initial point x', y', z' and the components of a final ray σ, τ, v, we know the values of the partial derivatives of W; hence (6·10) give us the components of the initial ray and (6·11) establish connections between the coordinates of any point on the final ray. *In fact,* (6·11) *are the equations of the final rays.* In particular, if there is a source of light at x', y', z', the congruence of final rays is given by (6·11), σ, τ, v taking arbitrary values subject to (6·2).

Let us now consider how W is to be calculated. We shall, however, confine ourselves to the case of an instrument involving only one reflection or refraction, because the extension of the method to the case of a general instrument simply requires the combination of the reasoning now to be given with that already given for V in § 5.

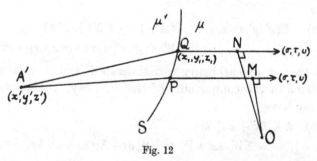

Fig. 12

In Fig. 12 $A'(x', y', z')$ is a given point: QN is a directed line in

† In certain special cases these five infinitesimals are not independent: this will occur when the congruence of rays from a point B', chosen arbitrarily in the neighbourhood of A', fails to give final rays having all directions adjacent to the direction σ, τ. We shall, for simplicity, exclude such cases from consideration.

‡ The answering of (e) merely involves the interchange of initial and final media.

the final medium with assigned components σ, τ, v, but $A'QN$ is not a natural ray in general, the law of refraction or reflection not being satisfied at Q. We can, however, express as a function of x', y', z', x_1, y_1, z_1 (the coordinates of Q) and σ, τ, v the optical length $[A'QN]$, where N is the foot of the perpendicular from the origin O on QN. It is in fact

$$(6\cdot12) \qquad L = [A'QN] = \mu'(\Sigma(x'-x_1)^2)^{\frac{1}{2}} - \Sigma\sigma x_1.$$

Now let $A'PM$ be a natural ray, the components of PM being σ, τ, v and M being the foot of the perpendicular from O. The plane OMN is perpendicular to the common direction of PM and QN, and hence if PQ is infinitesimal, MN is an infinitesimal displacement on the normal surface through M to the final rays of the congruence from a source at A'. Therefore to the first order

$$(6\cdot13) \qquad [A'QN] = [A'QM] = [A'PM].$$

Thus, letting Q tend to coincidence with P, we see that the coordinates x_1, y_1, z_1 at P are such that for an arbitrary displacement δx_1, δy_1, δz_1 on the surface S we have $\delta L = 0$. Therefore if the equation of S is

$$(6\cdot14) \qquad f(x_1, y_1, z_1) = 0,$$

we have for the natural ray

$$(6\cdot15) \qquad \frac{\partial L}{\partial x_1} = \lambda\frac{\partial f}{\partial x_1}, \quad \frac{\partial L}{\partial y_1} = \lambda\frac{\partial f}{\partial y_1}, \quad \frac{\partial L}{\partial z_1} = \lambda\frac{\partial f}{\partial z_1},$$

where λ is undetermined. Explicitly we have

$$(6\cdot16) \qquad \begin{cases} -\dfrac{\mu'(x'-x_1)}{\rho'} - \sigma = \lambda\dfrac{\partial f}{\partial x_1}, \\[2ex] -\dfrac{\mu'(y'-y_1)}{\rho'} - \tau = \lambda\dfrac{\partial f}{\partial y_1}, \\[2ex] -\dfrac{\mu'(z'-z_1)}{\rho'} - v = \lambda\dfrac{\partial f}{\partial z_1}, \end{cases}$$

where

$$(6\cdot17) \qquad \rho'^2 = \Sigma(x'-x_1)^2.$$

If we eliminate λ, solve for x_1, y_1, z_1 from $(6\cdot14)$ and $(6\cdot16)$, and substitute in $(6\cdot12)$, we have the function W,

(6·18) $W(x', y', z', \sigma, \tau) = L,$

(6·2) being used to express v in terms of σ, τ.

The calculation of W is generally difficult, but not as difficult as that of V. Let us calculate W for *refraction through a plane*. Let the plane $z = 0$ be the plane of refraction, separating the initial medium $z < 0$ of index μ' from the final medium $z > 0$ of index μ (Fig. 13).

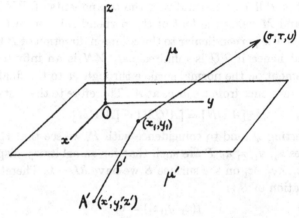

Fig. 13

Since $z_1 = 0$, the optical length L of (6·12) is

(6·19) $L = \mu'\{(x'-x_1)^2 + (y'-y_1)^2 + z'^2\}^{\frac{1}{2}} - \sigma x_1 - \tau y_1,$

and this is to be made stationary for arbitrary variations of x_1, y_1, so that in consequence

(6·20) $\begin{cases} \dfrac{\mu'(x'-x_1)}{\rho'} + \sigma = 0, \quad \dfrac{\mu'(y'-y_1)}{\rho'} + \tau = 0, \\ \rho' = \{(x'-x_1)^2 + (y'-y_1)^2 + z'^2\}^{\frac{1}{2}}. \end{cases}$

Thus

(6·21) $\begin{cases} \sigma^2 + \tau^2 = \dfrac{\mu'^2(\rho'^2 - z'^2)}{\rho'^2}, \quad \rho'^2 = \dfrac{\mu'^2 z'^2}{\mu'^2 - \sigma^2 - \tau^2}, \\ \rho' = -\dfrac{\mu' z'}{(\mu'^2 - \sigma^2 - \tau^2)^{\frac{1}{2}}}, \end{cases}$

the negative sign being taken in the last expression because $\rho' > 0$, $z' < 0$.

Then

(6·22) $x_1 = x' + \sigma\rho'/\mu'$, $y_1 = y' + \tau\rho'/\mu'$,

$\sigma x_1 + \tau y_1 = \sigma x' + \tau y' + (\sigma^2 + \tau^2)\rho'/\mu'$,

and so

(6·23) $W = L = \mu'\rho' - \sigma x_1 - \tau y_1$

$$= -\frac{\mu'^2 z'}{(\mu'^2 - \sigma^2 - \tau^2)^{\frac{1}{2}}} - \sigma x' - \tau y' + \frac{z'(\sigma^2 + \tau^2)}{(\mu'^2 - \sigma^2 - \tau^2)^{\frac{1}{2}}}$$

$$= -\sigma x' - \tau y' - z'(\mu'^2 - \sigma^2 - \tau^2)^{\frac{1}{2}}.$$

This is the characteristic function W for refraction across the plane $z = 0$.

The equations (6·11) give as the equations for the final rays

(6·24) $\begin{cases} x - z\dfrac{\sigma}{v} = x' - \dfrac{z'\sigma}{(\mu'^2 - \sigma^2 - \tau^2)^{\frac{1}{2}}}, \\[3mm] y - z\dfrac{\tau}{v} = y' - \dfrac{z'\tau}{(\mu'^2 - \sigma^2 - \tau^2)^{\frac{1}{2}}}; \end{cases}$

since $v^2 = \mu^2 - \sigma^2 - \tau^2$, these equations may be written

(6·25) $$\frac{x - x'}{\sigma} = \frac{y - y'}{\tau} = \frac{z}{(\mu^2 - \sigma^2 - \tau^2)^{\frac{1}{2}}} - \frac{z'}{(\mu'^2 - \sigma^2 - \tau^2)^{\frac{1}{2}}}.$$

If $A'(x', y', z')$ is a point source of light, we observe that a final ray with components σ, τ cuts the normal from A' to the refracting plane at

(6·26) $x = x'$, $y = y'$, $z = z'\dfrac{(\mu^2 - \sigma^2 - \tau^2)^{\frac{1}{2}}}{(\mu'^2 - \sigma^2 - \tau^2)^{\frac{1}{2}}};$

the value of z may also be written

(6·27) $$z = z'\frac{\mu\gamma}{\sqrt{\mu'^2 - \mu^2(1 - \gamma^2)}},$$

a value easily checked by an elementary argument based on the law of refraction (2·18).

7. The characteristic function T.

Let N', N be the feet of the perpendiculars dropped from the origin O on the initial and final portions of a ray passing through an instrument (Fig. 14).† We define T by

(7·1) $T = [N'N]$.

† As for V and W, we might employ two systems of coordinates, but we shall not do so.

We shall confine our attention to those instruments for which the directions of the initial and final portions of a ray define the ray completely, or at most define a finite number of rays. This will not be the case, for example, if the instrument is a cylindrical mirror, because then a set of parallel rays incident along a

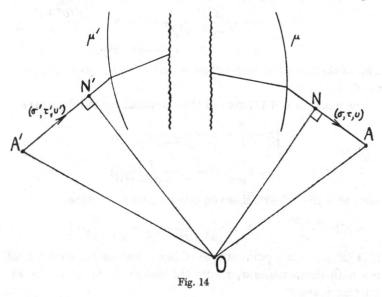

Fig. 14

generator give rise to a set of parallel final rays, and so the initial and final directions do not define a complete ray. The same applies to the case of refraction at a cylindrical surface, and, more generally, to reflection or refraction at a developable surface. Such instruments will not be considered.

Since by hypothesis the components σ', τ', υ', σ, τ, υ determine a complete ray, it is evident that T is a function of σ', τ', σ, τ, since υ', υ are given in terms of these four quantities by the identities (5·13). Thus we may write

(7·2) $$T = T(\sigma', \tau', \sigma, \tau).$$

T is to be regarded as a third *characteristic function* of the instrument; it may be called the *angle-characteristic*. The continuation process employed for V and W is available, and it is not necessary that N' and N should lie respectively on initial and final rays: they may lie on these rays produced. We shall now show the connections between T and V and W. If $A'(x',y',z')$ and $A(x,y,z)$ are any points on the initial and final rays respectively, we have

(7·3) $V = [A'A] = [A'N']+[N'N]+[NA],$

or

(7·4) $V(x',y',z',x,y,z)$
$$= T(\sigma',\tau',\sigma,\tau) - (\sigma'x'+\tau'y'+v'z') + (\sigma x+\tau y+vz),$$

or, by (6·5),

(7·5) $W(x',y',z',\sigma,\tau) = T(\sigma',\tau',\sigma,\tau) - (\sigma'x'+\tau'y'+v'z').$

Let us now give arbitrary variations to A' and A, with consequent variations in the components of the initial and final rays. Differentiation of (7·4) gives

(7·6) $\Sigma V_{x'}\,\delta x' + \Sigma V_x\,\delta x = T_{\sigma'}\,\delta\sigma' + T_{\tau'}\,\delta\tau' + T_\sigma\,\delta\sigma + T_\tau\,\delta\tau$
$$- \Sigma\sigma'\,\delta x' - \Sigma x'\,\delta\sigma' + \Sigma\sigma\,\delta x + \Sigma x\,\delta\sigma,$$

the subscripts as usual denoting partial differentiation; hence, by (5·12) and (6·8),

(7·7) $(T_{\sigma'}-x'+z'\sigma'/v')\,\delta\sigma' + (T_{\tau'}-y'+z'\tau'/v')\,\delta\tau'$
$$+ (T_\sigma+x-z\sigma/v)\,\delta\sigma + (T_\tau+y-z\tau/v)\,\delta\tau = 0.$$

But the four differentials occurring here may be regarded as arbitrary and independent; therefore

(7·8) $x'-z'\sigma'/v' = T_{\sigma'},\quad y'-z'\tau'/v' = T_{\tau'},$

(7·9) $x-z\sigma/v = -T_\sigma,\quad y-z\tau/v = -T_\tau.$

The function T being supposed known, (7·8) are *the equations of initial rays* and (7·9) *the equations of final rays*. Thus a knowledge of T provides us with an answer to the question (f) raised in § 5.

Let us now see how the function T is to be calculated for a given instrument, starting with a simple instrument in which only one reflection or refraction at a surface S is involved.

Let $\sigma',\tau',\sigma,\tau$, the initial and final components, be assigned, and let $N'QN$ be a broken line, $N'Q$ having components σ',τ' and QN components σ,τ and N',N being the feet of perpendiculars dropped from the origin O (Fig. 15). If x, y, z are the coordinates of Q, we have for the optical length

(7·10) $$[N'QN] = \Sigma\sigma'x - \Sigma\sigma x.$$

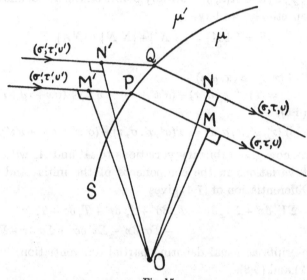

Fig. 15

Now let $M'PM$ be the natural ray with the assigned components, M', M being the feet of perpendiculars from O, and let Q be adjacent to P. It is easily seen, as in § 6, from Fermat's principle, that to the first order

(7·11) $$[N'QN] = [M'PM].$$

In fact, for arbitrary variations of Q on the surface S, $[N'QN]$ has a stationary value for the natural ray. Thus, by (7·10),

(7·12) $$\Sigma(\sigma' - \sigma)\,\delta x = 0$$

for all variations δx, δy, δz on S. Consequently, the vector with components

(7·13) $$\sigma - \sigma', \quad \tau - \tau', \quad \upsilon - \upsilon'$$

is normal to S at the point of incidence, as indeed we already knew from (2·13) for reflection and from (2·17) for refraction.

The mode of evaluation of T depends on the analytical form in which it is convenient to represent the surface. We have, by (7·10),

(7·14) $T(\sigma', \tau', \sigma, \tau) = (\sigma' - \sigma)x + (\tau' - \tau)y + (v' - v)z,$

from which x, y, z are to be eliminated by the stationary property (7·12).

If the surface S is given in the form

(7·15) $F(x, y, z) = 0,$

then (7·12) tells us that

(7·16) $\sigma - \sigma' = \lambda F_x, \quad \tau - \tau' = \lambda F_y, \quad v - v' = \lambda F_z,$

where λ is undetermined, and the subscripts indicate partial derivatives. Our procedure then is to eliminate λ, x, y, z from the five equations (7·14), (7·15), (7·16): we are also to use (5·13) to eliminate v' and v.

If the equation of S is given in the form

(7·17) $z = f(x, y),$

we know that the direction cosines of the normal have the ratios

(7·18) $f_x : f_y : -1,$

where f_x, f_y are partial derivatives. Hence

(7·19) $\dfrac{\sigma - \sigma'}{v - v'} = -f_x, \quad \dfrac{\tau - \tau'}{v - v'} = -f_y.$

If these two equations are solved for x, y, and z then found from (7·17), we may substitute in (7·14), and so by means of (5·13) obtain T as a function of the required arguments $\sigma', \tau', \sigma, \tau$.

There is yet a third method, dependent on a knowledge of the tangential equation of the surface S. Let l, m, n be the direction cosines of the normal to S at the point of incidence, and let p be the perpendicular distance from the origin O to the tangent plane to S at this point. The tangential equation of S is then of the form

(7·20) $p = \Phi(l, m, n);$

since $l^2 + m^2 + n^2 = 1$ this may be written in the form

(7·21) $p = \phi\left(\dfrac{l}{n}, \dfrac{m}{n}\right).$

But

(7·22) $\sigma - \sigma' = \lambda l, \quad \tau - \tau' = \lambda m, \quad v - v' = \lambda n,$

where λ is undetermined. Then (7·14) gives

(7·23) $T = -\lambda(lx + my + nz) = -\lambda p.$

But from (7·21) and (7·22)

(7·24) $\begin{cases} p = \phi\left(\dfrac{\sigma - \sigma'}{v - v'}, \dfrac{\tau - \tau'}{v - v'}\right), \\[2mm] \lambda^2 = (\sigma - \sigma')^2 + (\tau - \tau')^2 + (v - v')^2. \end{cases}$

Therefore

(7·25) $T = \pm\left[(\sigma - \sigma')^2 + (\tau - \tau')^2 + (v - v')^2\right]^{\frac{1}{2}} \phi\left(\dfrac{\sigma - \sigma'}{v - v'}, \dfrac{\tau - \tau'}{v - v'}\right),$

from which v', v are to be eliminated by (5·13). The explicit ambiguity in sign and those implicit in v' and v are to be removed by inspection in any particular case.

Let us now consider the calculation of T for a general instrument formed of any number of media. We shall use the notation of Fig. 8, § 5, and put

$\sigma', \tau', v' =$ components of initial ray,

$\sigma, \tau, v =$ components of final ray,

$\sigma_i, \tau_i, v_i =$ components of ray in medium M_i of index μ_i

$$(i = 1, 2, ..., n - 1).$$

Any two consecutive media may be regarded as forming an instrument. Let $T_{i-1,i}$ be the characteristic function for the instrument formed by the media M_{i-1}, M_i. Now a ray traversing M_i may be regarded either as a *final* ray for the combination M_{i-1}, M_i or as an *initial* ray for the combination M_i, M_{i+1}. If x_i, y_i, z_i are the coordinates of any point on a ray in M_i, we have then, by (7·8), (7·9),

(7·26) $\begin{cases} x_i - z_i \dfrac{\sigma_i}{v_i} = -\dfrac{\partial}{\partial \sigma_i} T_{i-1,i}, \quad y_i - z_i \dfrac{\tau_i}{v_i} = -\dfrac{\partial}{\partial \tau_i} T_{i-1,i}, \\[3mm] x_i - z_i \dfrac{\sigma_i}{v_i} = \dfrac{\partial}{\partial \sigma_i} T_{i,i+1}, \quad y_i - z_i \dfrac{\tau_i}{v_i} = \dfrac{\partial}{\partial \tau_i} T_{i,i+1}. \end{cases}$

Hence by subtraction

(7·27) $\qquad \dfrac{\partial}{\partial \sigma_i}(T_{i-1,i}+T_{i,i+1})=0, \qquad \dfrac{\partial}{\partial \tau_i}(T_{i-1,i}+T_{i,i+1})=0,$

these equations holding for $i = 1, 2, \ldots, n-1$, the subscripts 0 and n being attached to the initial and final media respectively. Consider now a ray traversing the complete instrument. From its definition as an optical length, it follows that T for the whole instrument is the sum of these functions for the simple instruments formed from pairs of consecutive media, that is,

(7·28) $\qquad\qquad T = T_{0,1}+T_{1,2}+\ldots+T_{n-1,n}.$

Now in the functions on the right there are involved all the components σ_i, τ_i for the rays in all the media. But any particular

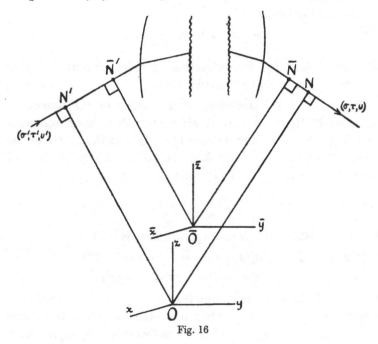

Fig. 16

pair σ_i, τ_i enter only into $T_{i-1,i}$ and $T_{i,i+1}$. Hence it follows from (7·27) that T as given by (7·28) has a stationary value with respect to arbitrary variations of the intermediate components. Thus we have the following rule for the calculation of T: *Find the characteristic function for each pair of adjacent media and add together the results. Eliminate the intermediate components by means of the equations which express the fact that the sum so obtained has a stationary value with respect to arbitrary variations of the intermediate components.*

The T-function of a system will change when the axes of coordinates are changed. Let $Oxyz$, $\bar{O}\bar{x}\bar{y}\bar{z}$ be two sets of parallel axes, the point $\bar{O}$ having coordinates a, b, c relative to $Oxyz$. Denoting by T, $\bar{T}$ the functions for these two systems of axes, we have (Fig. 16)

$$(7\cdot29) \qquad T = [N'N], \quad \bar{T} = [\bar{N}'\bar{N}],$$

and hence, since $N'\bar{N}'$, $N\bar{N}$ are the projections of $O\bar{O}$ on the initial and final rays,

$$(7\cdot30) \qquad \bar{T} = T + a(\sigma - \sigma') + b(\tau - \tau') + c(v - v').$$

8. The T-function for reflection or refraction at a sphere or a paraboloid of revolution.

The method of calculating T by means of the tangential equation $(7\cdot21)$ is convenient when the reflecting or refracting surface is spherical. Let us take the origin at the centre of the sphere: then $p = R$, where R is the radius of the sphere, and hence the function ϕ of $(7\cdot21)$ is simply a constant,

$$(8\cdot1) \qquad \phi\left(\frac{l}{n}, \frac{m}{n}\right) = R.$$

Hence by $(7\cdot25)$ *the T-function for a sphere is*

$$(8\cdot2) \qquad \begin{aligned} T &= \pm R[(\sigma - \sigma')^2 + (\tau - \tau')^2 + (v - v')^2]^{\frac{1}{2}} \\ &= \pm R[\mu^2 + \mu'^2 - 2(\sigma\sigma' + \tau\tau' + vv')]^{\frac{1}{2}}, \end{aligned}$$

from which the ambiguous sign is to be removed by special considerations. Should we wish to remove the origin from the centre of the sphere, we may use $(7\cdot30)$. The formula $(8\cdot2)$ applies both to reflection and to refraction.

To show how all ambiguities of sign are to be removed, let us consider internal reflection (Fig. 17a) and external reflection

(Fig. 17b). By consideration of the optical path, it is clear that T is positive in the former case and negative in the latter. This determines the sign in (8·2). Moreover, the axes being as shown, we have (the *positive* square root being indicated in each case)

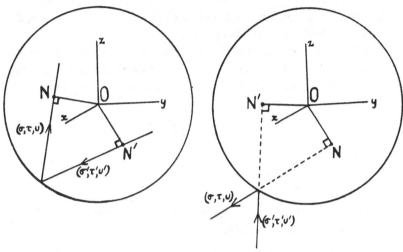

Fig. 17a Fig. 17b

(8·3) Fig. 17a:
$$0 > v' = -(\mu'^2 - \sigma'^2 - \tau'^2)^{\frac{1}{2}} = -\mu'(1 - \alpha'^2 - \beta'^2)^{\frac{1}{2}},$$
$$0 < v = (\mu^2 - \sigma^2 - \tau^2)^{\frac{1}{2}} = \mu(1 - \alpha^2 - \beta^2)^{\frac{1}{2}};$$

(8·4) Fig. 17b:
$$0 < v' = (\mu'^2 - \sigma'^2 - \tau'^2)^{\frac{1}{2}} = \mu'(1 - \alpha'^2 - \beta'^2)^{\frac{1}{2}},$$
$$0 > v = -(\mu^2 - \sigma^2 - \tau^2)^{\frac{1}{2}} = -\mu(1 - \alpha^2 - \beta^2)^{\frac{1}{2}};$$

where, of course, $\mu' = \mu$, this being the refractive index for the medium in which the rays lie. Thus, with all ambiguities removed, we have

(8·5) Fig. 17a:
$$T = R[2\mu^2 - 2(\sigma\sigma' + \tau\tau' + vv')]^{\frac{1}{2}}$$
$$= R\sqrt{2}[\mu^2 - \sigma\sigma' - \tau\tau' + (\mu^2 - \sigma^2 - \tau^2)^{\frac{1}{2}}(\mu^2 - \sigma'^2 - \tau'^2)^{\frac{1}{2}}]^{\frac{1}{2}}$$
$$= \mu R\sqrt{2}[1 - \alpha\alpha' - \beta\beta' + (1 - \alpha^2 - \beta^2)^{\frac{1}{2}}(1 - \alpha'^2 - \beta'^2)^{\frac{1}{2}}]^{\frac{1}{2}};$$

(8·6)   Fig. 17b:

$$T = - R[2\mu^2 - 2(\sigma\sigma' + \tau\tau' + \upsilon\upsilon')]^{\frac{1}{2}}$$

$$= - R\sqrt{2}[\mu^2 - \sigma\sigma' - \tau\tau' + (\mu^2 - \sigma^2 - \tau^2)^{\frac{1}{2}} (\mu^2 - \sigma'^2 - \tau'^2)^{\frac{1}{2}}]^{\frac{1}{2}}$$

$$= - \mu R\sqrt{2}[1 - \alpha\alpha' - \beta\beta' + (1 - \alpha^2 - \beta^2)^{\frac{1}{2}} (1 - \alpha'^2 - \beta'^2)^{\frac{1}{2}}]^{\frac{1}{2}}.$$

As usual, α', β' are direction cosines of the incident ray, and α, β direction cosines of the reflected ray.

Let us now consider the case where the reflecting or refracting surface is a paraboloid of revolution. Let us take the origin at the vertex and the z-axis along the axis of revolution. Then the equation of the surface is

(8·7) $$z = \frac{1}{2R} (x^2 + y^2),$$

R being the radius of curvature at the vertex.

By (7·14) we have

(8·8) $$T = (\sigma' - \sigma) x + (\tau' - \tau) y + (\upsilon' - \upsilon) z,$$

where this is to have a stationary value with respect to variations of x, y, z on the surface. Thus

(8·9) $$\sigma' - \sigma + (\upsilon' - \upsilon)\frac{\partial z}{\partial x} = 0, \quad \tau' - \tau + (\upsilon' - \upsilon)\frac{\partial z}{\partial y} = 0.$$

But by (8·7)

(8·10) $$\frac{\partial z}{\partial x} = \frac{x}{R}, \quad \frac{\partial z}{\partial y} = \frac{y}{R};$$

hence

(8·11) $$\begin{cases} x = - R\dfrac{\sigma' - \sigma}{\upsilon' - \upsilon}, \quad y = - R\dfrac{\tau' - \tau}{\upsilon' - \upsilon}, \\ z = \tfrac{1}{2}R\dfrac{(\sigma' - \sigma)^2 + (\tau' - \tau)^2}{(\upsilon' - \upsilon)^2}. \end{cases}$$

Substitution in (8·8) gives as *the T-function for a paraboloid of revolution*

(8·12) $$T = -\tfrac{1}{2}R\frac{(\sigma' - \sigma)^2 + (\tau' - \tau)^2}{\upsilon' - \upsilon}.$$

For a *mirror* in the form of a paraboloid of revolution, we have $\mu' = \mu$, and hence

(8·13) $$T = -\tfrac{1}{2}\mu R\frac{(\alpha' - \alpha)^2 + (\beta' - \beta)^2}{\gamma' - \gamma}.$$

Ambiguities of sign are intrinsic here on account of the dual signs in the expressions for γ', γ in terms of the other direction cosines. If the incident rays travel in the negative sense of the z-axis, and the reflected rays in the positive sense, we have

$$(8\cdot14) \qquad \gamma' = -(1-\alpha'^2-\beta'^2)^{\frac{1}{2}}, \quad \gamma = (1-\alpha^2-\beta^2)^{\frac{1}{2}},$$

and substitution in (8·13) gives T *for a paraboloidal mirror* free from ambiguity as

$$(8\cdot15) \qquad T = \tfrac{1}{2}\mu R \frac{(\alpha'-\alpha)^2+(\beta'-\beta)^2}{(1-\alpha'^2-\beta'^2)^{\frac{1}{2}}+(1-\alpha^2-\beta^2)^{\frac{1}{2}}}.$$

The T-function is most useful for the discussion of final rays when the incident rays are parallel to one another. The final rays are, as in (7·9),

$$(8\cdot16) \qquad x-z\frac{\sigma}{v} = -T_\sigma, \quad y-z\frac{\tau}{v} = -T_\tau,$$

and in the functions on the right σ', τ' are now to be considered as constants.

We may easily verify from (8·12) the fact that when the incident rays are all parallel to the axis of a paraboloidal mirror, the reflected rays pass through the focus. We have $\sigma' = \tau' = 0$, $v' = -\mu$, and it is a matter of indifference whether we make these substitutions before or after differentiating with respect to σ and τ. Thus we may put from (8·12)

$$(8\cdot17) \qquad T = \tfrac{1}{2}R\frac{\sigma^2+\tau^2}{\mu+v} = \tfrac{1}{2}R(\mu-v),$$

$$T_\sigma = -\tfrac{1}{2}R\frac{\partial v}{\partial \sigma} = \tfrac{1}{2}R\frac{\sigma}{v}, \quad T_\tau = \tfrac{1}{2}R\frac{\tau}{v},$$

and (8·16) read

$$(8\cdot18) \qquad x = \frac{\sigma}{v}(z-\tfrac{1}{2}R), \quad y = \frac{\tau}{v}(z-\tfrac{1}{2}R),$$

which equations show that all the reflected rays pass accurately through the point $(0,0,\tfrac{1}{2}R)$.

The above expressions for T are exact: we shall consider in Chapter IV approximate forms of T, useful when we have to deal with a small bundle of incident rays.

CHAPTER III

THIN BUNDLES OF RAYS

9. Foci and focal lines.

Let us consider the final congruence of rays formed by the passage of light through an instrument from a source at a point x', y', z'. The equations of the final rays are, as in (6·11),

$$(9·1) \qquad x - z\sigma/v = -W_\sigma, \quad y - z\tau/v = -W_\tau;$$

x', y', z' are involved in W, but they are to be treated as constants in the present work. If the initial rays do not come from a point source at finite distance, but are parallel, we use the T-function instead of W, its arguments σ', τ' being constants. As a matter of fact, we are really concerned at present only with the final congruence of rays. We are indifferent as to its origin. We know that any normal congruence can be produced by reflection at a suitable mirror of light emanating from an assigned point source. Consequently, as far as the study of the general properties of a normal congruence of rays is concerned, (9·1) are completely general, or, indeed, similar equations with T instead of W.

If, on a given ray R, there is a point P such that some rays, making infinitesimal angles with R, cut R at P to the first order, then P is said to be a *focus*. Two lines are said to cut *to the first order* when their distance apart is an infinitesimal of order higher than the first.

Suppose now that a ray R and an adjacent ray cut to the first order at x, y, z. We shall have equations of the form (9·1) satisfied by each ray, x, y, z being (to the first order) the same for both rays, but the components σ, τ differing infinitesimally. We may in fact differentiate (9·1), putting

$$(9·2) \qquad \delta x = \delta y = \delta z = 0.$$

Hence

$$(9·3) \qquad -z\,\delta(\sigma/v) = -\delta W_\sigma, \quad -z\,\delta(\tau/v) = -\delta W_\tau,$$

or, since $\qquad \delta v = -(\sigma\,\delta\sigma + \tau\,\delta\tau)/v,$

$$(9\cdot4) \quad \begin{cases} \{z(\sigma^2+v^2)/v^3 - W_{\sigma\sigma}\}\,\delta\sigma + \{z\sigma\tau/v^3 - W_{\sigma\tau}\}\,\delta\tau = 0, \\ \{z\sigma\tau/v^3 - W_{\sigma\tau}\}\,\delta\sigma + \{z(\tau^2+v^2)/v^3 - W_{\tau\tau}\}\,\delta\tau = 0. \end{cases}$$

Elimination of $\delta\sigma$, $\delta\tau$ gives

$$(9\cdot5) \quad \{z(\sigma^2+v^2)/v^3 - W_{\sigma\sigma}\}\{z(\tau^2+v^2)/v^3 - W_{\tau\tau}\}$$
$$-\{z\sigma\tau/v^3 - W_{\sigma\tau}\}^2 = 0,$$

a quadratic equation for z. Since, as will be shown below, the roots are real, *this equation, with* $(9\cdot1)$, *determines two foci on the given ray.*

To show that the roots are necessarily real, we take special axes, the z-axis being coincident with the ray R, for which then we have

$$(9\cdot6) \qquad\qquad \sigma = \tau = 0, \quad v = \mu.$$

The partial derivatives of W occurring above were to be evaluated for the values of σ, τ belonging to the ray R: for our special choice of axes, these are as in $(9\cdot6)$. Since $(9\cdot1)$ are to be satisfied by $x = y = \sigma = \tau = 0$, we have

$$(9\cdot7) \qquad\qquad W_\sigma = W_\tau = 0$$

for $\sigma = \tau = 0$.

We are still free to rotate the axes about the ray R. Rotating through an angle θ gives a transformation

$$(9\cdot8) \quad \bar{x} = x\cos\theta + y\sin\theta, \quad \bar{y} = -x\sin\theta + y\cos\theta, \quad \bar{z} = z,$$

and hence, since σ, τ, v differ from direction cosines only by a constant factor,

$$(9\cdot9) \quad \bar{\sigma} = \sigma\cos\theta + \tau\sin\theta, \quad \bar{\tau} = -\sigma\sin\theta + \tau\cos\theta, \quad \bar{v} = v.$$

Now W, as an optical length, has a value independent of the directions of the axes. Thus

$$(9\cdot10) \quad W_\sigma = W_{\bar\sigma}\frac{\partial\bar\sigma}{\partial\sigma} + W_{\bar\tau}\frac{\partial\bar\tau}{\partial\sigma}, \quad W_\tau = W_{\bar\sigma}\frac{\partial\bar\sigma}{\partial\tau} + W_{\bar\tau}\frac{\partial\bar\tau}{\partial\tau},$$

$$W_{\sigma\tau} = W_{\bar\sigma\bar\sigma}\frac{\partial\bar\sigma}{\partial\sigma}\frac{\partial\bar\sigma}{\partial\tau} + W_{\bar\sigma\bar\tau}\left(\frac{\partial\bar\sigma}{\partial\sigma}\frac{\partial\bar\tau}{\partial\tau} + \frac{\partial\bar\tau}{\partial\sigma}\frac{\partial\bar\sigma}{\partial\tau}\right) + W_{\bar\tau\bar\tau}\frac{\partial\bar\tau}{\partial\sigma}\frac{\partial\bar\tau}{\partial\tau}$$

$$= (W_{\bar\sigma\bar\sigma} - W_{\bar\tau\bar\tau})\sin\theta\cos\theta + W_{\bar\sigma\bar\tau}(\cos^2\theta - \sin^2\theta).$$

Thus, given the axes $\bar{x}$, $\bar{y}$ arbitrarily in a plane perpendicular to R, we have merely to choose the x, y axes so that

$$(9 \cdot 11) \qquad \tan 2\theta = -\frac{2W_{\sigma\tau}}{W_{\sigma\sigma} - W_{\tau\tau}}$$

in order to make

$$(9 \cdot 12) \qquad W_{\sigma\tau} = 0$$

for the ray R.

With this special choice of axes, for which (9·7) and (9·12) are satisfied, the equations (9·4) reduce to

$$(9 \cdot 13) \qquad (z/\mu - W_{\sigma\sigma})\,\delta\sigma = 0, \qquad (z/\mu - W_{\tau\tau})\,\delta\tau = 0,$$

and the quadratic (9·5) reduces to

$$(9 \cdot 14) \qquad (z/\mu - W_{\sigma\sigma})(z/\mu - W_{\tau\tau}) = 0.$$

The roots are therefore real, namely,

$$(9 \cdot 15) \qquad z_1 = \mu W_{\sigma\sigma}, \quad z_2 = \mu W_{\tau\tau}.$$

Avoiding the particular case where $W_{\sigma\sigma} = W_{\tau\tau}$, we see that (9·13) has the two solutions

$$(9 \cdot 16) \qquad z = z_1, \; \delta\tau = 0; \quad z = z_2, \; \delta\sigma = 0.$$

Thus each ray of a normal congruence possesses two foci, whose coordinates are given by (9·5) and (9·1) for general axes, and by (9·15) for the special axes.

By (9·1), (9·7) and (9·12), the equations of a general ray adjacent to R, referred to the special axes, are to the first order

$$(9 \cdot 17) \quad x - z\,\delta\sigma/\mu = -W_{\sigma\sigma}\,\delta\sigma, \quad y - z\,\delta\tau/\mu = -W_{\tau\tau}\,\delta\tau,$$

where $\delta\sigma$, $\delta\tau$ are the components of the ray and the partial derivatives are evaluated for $\sigma = \tau = 0$. By (9·15) these may be written

$$(9 \cdot 18) \qquad x = (z - z_1)\,\delta\sigma/\mu, \quad y = (z - z_2)\,\delta\tau/\mu.$$

All these rays, for arbitrary $\delta\sigma$, $\delta\tau$, cut the plane $z = z_1$ in the line

$$(9 \cdot 19) \qquad x = 0, \quad z = z_1,$$

and the plane $z = z_2$ in the line

$$(9 \cdot 20) \qquad y = 0, \quad z = z_2.$$

These lines (9·19), (9·20) (see Fig. 18) are called the *focal lines*. We have the following result: *All rays adjacent to any ray R cut*

*(to the first order) two focal lines, one through each focus of R, which
are perpendicular to one another and to R.*

This gives a very simple way of constructing an approximate
model of a thin bundle of a nor-
mal congruence when the focal
lines are known. We simply join
up all points of the two focal lines.

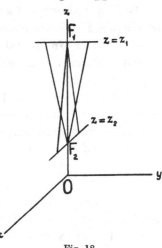

Let us now consider how the
focal lines are to be found for a
general system of axes of coordinates. We are to use (9·4), z
satisfying (9·5). Let the roots of
this equation be z_1, z_2, and let
the corresponding foci be F_1, F_2.
Let $\delta_1\sigma$, $\delta_1\tau$ be solutions of (9·4)
corresponding to $z = z_1$, and $\delta_2\sigma$,
$\delta_2\tau$ those corresponding to $z = z_2$.
Let l_1, m_1, n_1 be the direction

Fig. 18

cosines of the focal line at F_1 and l_2, m_2, n_2 those of the focal
line at F_2. It is evident, from consideration of the arrangement
of the rays shown in Fig. 18 for special axes, that the focal
line at F_1 is perpendicular to the directions with components
(σ, τ, υ) and $(\sigma + \delta_1\sigma, \tau + \delta_1\tau, \upsilon + \delta_1\upsilon)$. Hence

(9·21)
$$\begin{cases} l_1\sigma + m_1\tau + n_1\upsilon = 0, \\ l_1\delta_1\sigma + m_1\delta_1\tau + n_1\delta_1\upsilon = 0, \end{cases}$$

and so

(9·22) $l_1 : m_1 : n_1 = (\tau\delta_1\upsilon - \upsilon\delta_1\tau) : (\upsilon\delta_1\sigma - \sigma\delta_1\upsilon) : (\sigma\delta_1\tau - \tau\delta_1\sigma);$

similarly

(9·23) $l_2 : m_2 : n_2 = (\tau\delta_2\upsilon - \upsilon\delta_2\tau) : (\upsilon\delta_2\sigma - \sigma\delta_2\upsilon) : (\sigma\delta_2\tau - \tau\delta_2\sigma).$

*Equations (9·22) and (9·23) give the directions of the focal lines of
a ray with components σ, τ, υ, the axes being general. The ratios
$\delta_1\sigma/\delta_1\tau$ and $\delta_2\sigma/\delta_2\tau$ are to be found from (9·4), after inserting suc-
cessively the two values of z satisfying (9·5): $\delta_1\upsilon$ and $\delta_2\upsilon$ are to be
found from the identical relation written just above (9·4). We could
of course obtain the two values of $\delta\sigma/\delta\tau$ by solving the quadratic*

equation for this ratio obtained by eliminating z from (9·4); but we would not then know which focal line corresponded to which focus.

The focal properties of a normal congruence of rays may also be discussed by considering the congruence as the normals to a wave-surface. In general, two adjacent normals to a surface do not intersect to the first order: but if they are drawn from adjacent points on a line of curvature of the surface they do intersect to the first order. The two foci on each ray arise from the intersections of the ray with adjacent rays drawn from points on the two lines of curvature on the wave-surface.

A *pencil* of rays consists of a single infinity of rays: a pencil forms a ruled surface, which is *developable* if the adjacent rays on it intersect to the first order. It is evident that we can construct from a given normal congruence two singly-infinite sets of *developable pencils*, each developable pencil consisting of the rays which cut a wave-surface along a line of curvature. All the rays in each developable pencil touch a curve, called a *caustic curve*, which consists of the points of intersection of adjacent rays in the pencil; these points are centres of curvature of the wave-surface and foci on the rays, in the sense defined above. The totality of caustic curves form a *caustic surface* of two sheets, whose points are the centres of curvature of the wave-surface, or foci on the rays (two on each ray). All the rays touch the caustic surface, and a focal line of a given ray is simply a line through a focus, lying in the tangent plane to the caustic surface there and perpendicular to the given ray.

Let us now discuss *the variation in the cross-section of a thin bundle of rays*. Let us choose the z-axis along a ray of the bundle, and use the special axes for which (9·12) holds. The boundary of the bundle will be a pencil of rays, which may be defined by equations

$$(9·24) \qquad \sigma = f(u), \quad \tau = g(u),$$

where f, g are small functions of a parameter u. Now for any ray adjacent to the z-axis we have approximately from (9·1)

$$(9·25) \qquad x - z\sigma/\mu = -\sigma W_{\sigma\sigma}, \quad y - z\tau/\mu = -\tau W_{\tau\tau},$$

or, by (9·15),

(9·26) $x = \sigma(z - z_1)/\mu, \quad y = \tau(z - z_2)/\mu,$

where z_1, z_2 are the foci. The area of the section of the bundle by $z = $ const. is $\pm A$, where

(9·27) $A = \tfrac{1}{2}\int (x\,dy - y\,dx),$

taken round the bounding curve of the section in the sense of u increasing. By (9·26) this is

$$(9\cdot 28)\quad A = \frac{(z - z_1)(z - z_2)}{2\mu^2}\int (\sigma\,d\tau - \tau\,d\sigma)$$

$$= \frac{(z - z_1)(z - z_2)}{2\mu^2}\int (fg' - gf')\,du.$$

Between the foci $(z_1 - z)(z - z_2)$ is positive, and its greatest value is $\tfrac{1}{4}(z_1 - z_2)^2$, which occurs for $z = \tfrac{1}{2}(z_1 + z_2)$. The corresponding value of A is

$$(9\cdot 29)\qquad \bar{A} = -\frac{(z_1 - z_2)^2}{8\mu^2}\int (fg' - gf')\,du.$$

Therefore the area of any section is given in terms of this maximum area by

$$(9\cdot 30)\qquad A = 4\bar{A}\,\frac{(z_1 - z)(z - z_2)}{(z_1 - z_2)^2}.$$

The area of section is thus proportional to the product of the distances from the foci.

Since the arrangement of rays in a thin bundle is determined to the first order by one of its rays (called the central ray) and its focal lines, it follows that given a ray, incident on a given reflecting or refracting surface, and the focal lines of that ray, we should be able to find the focal lines of the reflected or refracted ray. This question will now be investigated in a special case, as an example of the use of the T-function.

A thin bundle of rays is reflected or refracted at a surface, the plane of incidence being a principal plane of curvature: the positions of the foci on the incident central ray are given, and one of its focal lines is parallel to the direction of a line of curvature on the reflecting or refracting surface at the point of incidence. It is required to find the focal lines of the reflected or refracted bundle.

Let us treat the case of refraction as the more general. Let us choose the origin at the point of incidence of the central ray, Oz along the normal to the refracting surface and Oxy in principal

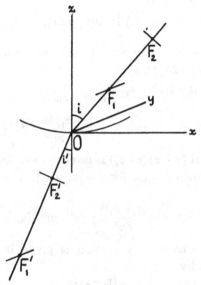

Fig. 19

directions of curvature. If the radii of curvature corresponding to Ox, Oy are R_1, R_2 respectively, the equation of the surface near the origin is approximately

$$(9\cdot31) \qquad z = \frac{x^2}{2R_1} + \frac{y^2}{2R_2}.$$

Let T be the characteristic function for the pair of media. Then, as in $(7\cdot14)$, $(7\cdot19)$,

$$(9\cdot32) \qquad T = (\sigma' - \sigma)\,x + (\tau' - \tau)\,y + (\upsilon' - \upsilon)\,z,$$

$$\frac{\sigma' - \sigma}{\upsilon' - \upsilon} = -\frac{x}{R_1}, \qquad \frac{\tau' - \tau}{\upsilon' - \upsilon} = -\frac{y}{R_2},$$

and so

$$(9\cdot33) \qquad T = -\frac{1}{2(\upsilon' - \upsilon)}\{R_1(\sigma' - \sigma)^2 + R_2(\tau' - \tau)^2\}.$$

This is of course only an approximate form, based on the approximate equation (9·31), and therefore valid only for rays incident near the origin. We note from (9·32) that $\sigma' - \sigma$, $\tau' - \tau$ are small: we may therefore, to our degree of approximation, substitute an approximate value for $v' - v$ in (9·33).

Let us suppose that the incident bundle has its central ray in the plane Oxz: the refracted bundle then also has its central ray in this plane, and we have, accurately for the central rays and approximately through the bundles,

$$(9·34) \qquad v' = \mu' \cos i', \quad v = \mu \cos i,$$

where μ', μ are the indices of refraction and i', i the angles of incidence and refraction respectively (Fig. 19). Accordingly we may write (9·33) in the form (accurate to the second order of small quantities)

$$(9·35) \qquad T = \tfrac{1}{2}k\{R_1(\sigma' - \sigma)^2 + R_2(\tau' - \tau)^2\},$$

$$k = \frac{1}{\mu \cos i - \mu' \cos i'}.$$

We note the following values for partial derivatives:

$$(9·36) \qquad \begin{cases} T_{\sigma'\sigma'} = T_{\sigma\sigma} = -T_{\sigma'\sigma} = kR_1, \\ T_{\tau'\tau'} = T_{\tau\tau} = -T_{\tau'\tau} = kR_2, \\ T_{\sigma\tau} = T_{\sigma\tau'} = T_{\sigma'\tau} = T_{\sigma'\tau'} = 0. \end{cases}$$

Now the equations of the incident and refracted rays are, as in (7·8), (7·9),

$$(9·37) \qquad \begin{cases} x' - z'\sigma'/v' = T_{\sigma'}, & y' - z'\tau'/v' = T_{\tau'}, \\ x - z\sigma/v = -T_{\sigma}, & y - z\tau/v = -T_{\tau}. \end{cases}$$

We may regard these as four equations connecting x, y, z, σ, τ when x', y', z', σ', τ' are given. If we regard z as also given, they determine x, y, σ, τ; in fact, they determine the final ray corresponding to any assigned initial ray.

Let the foci F'_1, F'_2 of the incident bundle be at $z = z'_1$, $z = z'_2$, and let the focal line at F'_1 be perpendicular to the plane of incidence (F'_1 is then called a *primary focus*), the focal line at F'_2 lying in the plane of incidence (F'_2 is then called a *secondary focus*).

Let us pass from the central ray to *any* adjacent ray. Differentiation of (9·37) gives, in view of (9·36),

$$(9·38) \quad \begin{cases} \delta x' - \delta(z'\sigma'/v') = kR_1(\delta\sigma' - \delta\sigma), \\ \qquad\qquad\quad \delta y' - \delta(z'\tau'/v') = kR_2(\delta\tau' - \delta\tau), \\ \delta x - \delta(z\sigma/v) = kR_1(\delta\sigma' - \delta\sigma), \\ \qquad\quad \delta y - \delta(z\tau/v) = kR_2(\delta\tau' - \delta\tau). \end{cases}$$

After differentiation we may insert the approximate values

$$(9·39) \quad \begin{cases} \sigma' = \mu'\sin i', \quad \tau' = 0, \quad v' = \mu'\cos i', \\ \sigma = \mu\sin i, \qquad \tau = 0, \quad v = \mu\cos i, \end{cases}$$

so that, by (6·8) and the corresponding accented equation,

$$(9·40) \quad \begin{cases} \delta v' = -\dfrac{\sigma'\,\delta\sigma'}{v'} = -\tan i'\,\delta\sigma', \quad \delta v = -\dfrac{\sigma\,\delta\sigma}{v} = -\tan i\,\delta\sigma, \\ \delta(\sigma'/v') = \dfrac{\delta\sigma'}{\mu'}\sec^3 i', \qquad\qquad \delta(\sigma/v) = \dfrac{\delta\sigma}{\mu}\sec^3 i. \end{cases}$$

Let the points (x', y', z'), $(x' + \delta x', y' + \delta y', z' + \delta z')$ coincide to the first order at the primary focus F_1', so that we have

$$(9·41) \qquad\qquad \delta x' = \delta y' = \delta z' = 0,$$

and consequently $\delta\tau' = 0$, since the incident ray must pass through the focal line at F_2'. Then (9·38) give

$$(9·42) \quad \begin{cases} -z_1'\delta(\sigma'/v') = kR_1(\delta\sigma' - \delta\sigma), \quad 0 = -kR_2\,\delta\tau, \\ \delta x - \delta(z\sigma/v) = kR_1(\delta\sigma' - \delta\sigma), \quad \delta y - \delta(z\tau/v) = -kR_2\,\delta\tau. \end{cases}$$

Hence $\delta\tau = 0$, showing that the varied refracted ray lies in the plane $y = 0$, as is indeed evident from symmetry. The refracted rays, corresponding to arbitrary $\delta\sigma'$, will all pass through a point (x, y, z) provided that the equations (9·42) can be satisfied with $\delta x = \delta y = \delta z = 0$. Thus we have to satisfy

$$(9·43) \quad \begin{cases} -\dfrac{z_1'}{\mu'}\sec^3 i'\,\delta\sigma' = kR_1(\delta\sigma' - \delta\sigma), \\ -\dfrac{z}{\mu}\sec^3 i\,\delta\sigma \;\;= kR_1(\delta\sigma' - \delta\sigma): \end{cases}$$

these give the focus z_1, on elimination of $\delta\sigma'$, $\delta\sigma$, by the formula

(9·44)
$$\frac{\mu \cos^3 i}{z_1} - \frac{\mu' \cos^3 i'}{z_1'} = \frac{1}{kR_1}:$$

it is a primary focus (F_1), since $\delta\tau = 0$.

To get the other focus, let the points

$$(x',y',z'), \quad (x'+\delta x', y'+\delta y', z'+\delta z')$$

coincide to the first order at the secondary focus F_2', so that (9·41) again hold. Since the ray must pass through the focal line at F_1', we have

(9·45)
$$\delta\sigma' = 0, \quad \delta v' = 0.$$

Putting $\delta x = \delta y = \delta z = 0$ in (9·38) to get a focus, and remembering (9·39), (9·40), we have to satisfy

(9·46)
$$\begin{cases} 0 = kR_1(-\delta\sigma), & -\dfrac{z_2'}{\mu'}\sec i'\,\delta\tau' = kR_2(\delta\tau' - \delta\tau), \\[2mm] -\dfrac{z}{\mu}\sec^3 i\,\delta\sigma = kR_1(-\delta\sigma), & -\dfrac{z}{\mu}\sec i\,\delta\tau = kR_2(\delta\tau' - \delta\tau): \end{cases}$$

these give the focus z_2, on elimination of $\delta\tau'$, $\delta\tau$, by the formula

(9·47)
$$\frac{\mu \cos i}{z_2} - \frac{\mu' \cos i'}{z_2'} = \frac{1}{kR_2}.$$

It is a secondary focus (F_2).

If ρ_1', ρ_2' denote the distances of the incident primary and secondary foci from O, and ρ_1, ρ_2 the distances of the refracted primary and secondary foci from O, all counted positive when measured in the sense of propagation, we have from (9·44), (9·47)

(9·48)
$$\begin{cases} \dfrac{\mu \cos^2 i}{\rho_1} - \dfrac{\mu' \cos^2 i'}{\rho_1'} = \dfrac{\mu \cos i - \mu' \cos i'}{R_1}, \\[3mm] \dfrac{\mu}{\rho_2} - \dfrac{\mu'}{\rho_2'} = \dfrac{\mu \cos i - \mu' \cos i'}{R_2}. \end{cases}$$

10. Aberration at a focal line.

The theory of focal lines developed in § 9 is an approximate first order theory. Let us now investigate more accurately the pattern formed by a bundle of rays on the plane passing through the focal line on a central ray, and perpendicular to the ray. The deviation of this pattern from the focal line is known as *aberration*.

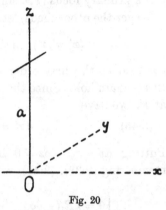

Fig. 20

Let us choose our axes so that the z-axis is the central ray, one focus being at the origin and the x-axis being a focal line. Using the W-function to describe the system of rays (supposed to emanate from a source x', y', z' in an initial medium), it follows from (9·1) and (9·4) that we have for the central ray ($\sigma = \tau = 0$)

$$(10\cdot1) \quad W_\sigma = 0, \quad W_\tau = 0, \quad W_{\sigma\tau} = 0, \quad W_{\tau\tau} = 0, \quad W_{\sigma\sigma} = a/\mu,$$

where $(0,0,a)$ is the other focus. We shall examine the aberration near the origin.

The exact equations of the rays are as in (9·1)

$$(10\cdot2) \qquad x - z\sigma/v = -W_\sigma, \quad y - z\tau/v = -W_\tau.$$

Expanding the right-hand sides in power series in σ, τ and putting $z = 0$, we see that a ray with direction cosines α, β, γ cuts the plane $z = 0$ at the point

$$(10\cdot3) \quad \begin{cases} x = -a\alpha + \tfrac{1}{2}(A\alpha^2 + 2B\alpha\beta + C\beta^2) + \dots, \\ y = \qquad \tfrac{1}{2}(B\alpha^2 + 2C\alpha\beta + D\beta^2) + \dots, \end{cases}$$

where

$$(10\cdot4) \quad \begin{cases} A = -\mu^2 W_{\sigma\sigma\sigma}, \quad B = -\mu^2 W_{\sigma\sigma\tau}, \\ C = -\mu^2 W_{\sigma\tau\tau}, \quad D = -\mu^2 W_{\tau\tau\tau}, \end{cases}$$

the partial derivatives being evaluated for $\sigma = \tau = 0$. Introducing spherical polar angles θ, ϕ to specify the direction of the ray, we have

$$(10\cdot5) \qquad \alpha = \sin\theta \cos\phi, \quad \beta = \sin\theta \sin\phi.$$

Including terms of the order of θ^2, but of no higher order, (10·3) give

$$(10\cdot6) \quad \begin{cases} x = -a\theta\cos\phi \\ \quad + \tfrac{1}{2}\theta^2(A\cos^2\phi + 2B\cos\phi\sin\phi + C\sin^2\phi), \\ y = \tfrac{1}{2}\theta^2(B\cos^2\phi + 2C\cos\phi\sin\phi + D\sin^2\phi). \end{cases}$$

To this order of approximation, it is sufficient to substitute in the expression for y approximate values for $\cos\phi$, $\sin\phi$ given by the first equation, namely,

$$(10\cdot7) \qquad \cos\phi = -\frac{x}{a\theta}, \quad \sin\phi = \pm\left(1 - \frac{x^2}{a^2\theta^2}\right)^{\frac{1}{2}}.$$

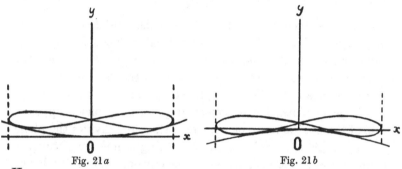

Fig. 21a   Fig. 21b

Hence

$$(10\cdot8) \quad 2a^2y = Bx^2 \pm 2Cx(a^2\theta^2 - x^2)^{\frac{1}{2}} + D(a^2\theta^2 - x^2).$$

If θ is held fixed, this is the equation of the locus of the points of intersection with the plane $z = 0$ of all rays making a small angle θ with the central ray: by varying θ we get the whole pattern.

The curve (10·8) with θ constant is obviously a flat curve near the x-axis. To make the radical real, we must take x so that

$$(10\cdot9) \qquad\qquad -a\theta \leqslant x \leqslant a\theta,$$

assuming a positive for simplicity. Thus the curve is bounded by the lines $x = \pm a\theta$, which it touches. Since two values of y correspond to each value of x except $x = 0$, $\pm a\theta$, the curve is a figure of eight. Two types are shown in Figs. 21a, b. For $x = \pm a\theta$, we have $y = \tfrac{1}{2}B\theta^2$, and for $x = 0$, $y = \tfrac{1}{2}D\theta^2$: at the latter point the slope is $\pm C\theta/a$. The curve cuts the x-axis if, and only if, $C^2 - BD > 0$.

To find the illumination produced on $z = 0$ by all rays adjacent to the central ray, we are to superimpose the curves for all small values of θ. Let us find the envelope of the curves. Differentiating (10·8) with respect to θ, we get, on division by $2a^2\theta$,

$$(10\cdot10) \qquad \pm \frac{Cx}{(a^2\theta^2 - x^2)^{\frac{1}{2}}} + D = 0,$$

and hence

$$(10\cdot11) \qquad (a^2\theta^2 - x^2)^{\frac{1}{2}} = \mp\, Cx/D.$$

Substitution in (10·8) gives for the envelope the parabola

$$(10\cdot12) \qquad y = \frac{BD - C^2}{2a^2D}\, x^2.$$

A mere reversal of the y-axis changes the sign of D, so that we may suppose $D > 0$ without loss of generality. Then the figures of eight cut the positive y-axis, as shown in Figs. 21a, b; Fig. 21a shows the case $BD - C^2 > 0$ and Fig. 21b the case $BD - C^2 < 0$. In either case the figures of eight lie entirely on one side of the parabolic envelope. No rays meet the plane $z = 0$ on the opposite side of the parabolic envelope, which therefore divides the plane into two regions—bright and dark—the light in the bright region being concentrated near the parabolic envelope. Thus the theory of geometrical optics indicates a sharp separation between light and darkness, but in reality the two regions will merge into one another with bright and dark diffraction bands.

11. Principal foci: aberration at a principal focus.

A point P on a ray R is said to be a *principal focus* if *all* rays which make with R an angle less than a small angle θ pass through P, distances of the order of θ^2 being neglected. A ray containing a principal focus is called a *principal ray*.

Using the function W, as at the beginning of § 9, to describe the congruence of rays—the T-function may of course be used similarly—the condition for a principal focus at x, y, z is obviously that the conditions of intersection (9·4) should be satisfied for arbitrary values of $\delta\sigma$, $\delta\tau$. Thus the equations

$$(11\cdot1) \quad z(\sigma^2 + v^2)/v^3 = W_{\sigma\sigma}, \quad z\sigma\tau/v^3 = W_{\sigma\tau}, \quad z(\tau^2 + v^2)/v^3 = W_{\tau\tau}$$

are to be satisfied. These are three equations for z, σ, τ, since v is given by

$$(11\cdot2) \qquad v^2 = \mu^2 - \sigma^2 - \tau^2.$$

When these quantities have been found, x and y are given by

$$(11\cdot3) \qquad x - z\sigma/v = -W_\sigma, \quad y - z\tau/v = -W_\tau.$$

Hence we can locate the principal foci (x, y, z) and the principal rays (σ, τ) through them when we know the function W. We note that the components of the principal rays satisfy

$$(11\cdot4) \qquad \frac{W_{\sigma\sigma}}{\mu^2 - \tau^2} = \frac{W_{\sigma\tau}}{\sigma\tau} = \frac{W_{\tau\tau}}{\mu^2 - \sigma^2}.$$

In general a given congruence of rays will possess a finite number of principal foci and principal rays.

If we take a principal ray for z-axis and the principal focus on it for origin, $(11\cdot1)$ and $(11\cdot3)$ must be satisfied with $x = y = z = 0$ Hence we have

$$(11\cdot5) \qquad W_\sigma = W_\tau = W_{\sigma\sigma} = W_{\sigma\tau} = W_{\tau\tau} = 0$$

for $\sigma = \tau = 0$.

Let us now investigate the pattern formed on the plane through a principal focus perpendicular to the principal ray. This plane is called a *focal plane*, and the deviation of the rays from the principal focus is known as *aberration*.

The exact equations of the rays are $(11\cdot3)$. Let us take the special axes of coordinates described above, so that the focal plane is $z = 0$. Developing the right-hand sides of $(11\cdot3)$ in power series, we see that the intersection of the ray with direction cosines α, β, γ with the focal plane is (on account of $(11\cdot5)$)

$$(11\cdot6) \qquad \begin{cases} x = \frac{1}{2}(A\alpha^2 + 2B\alpha\beta + C\beta^2) + \dots, \\ y = \frac{1}{2}(B\alpha^2 + 2C\alpha\beta + D\beta^2) + \dots, \end{cases}$$

where A, B, C, D are constants as given in $(10\cdot4)$. Introducing the angles θ, ϕ as in $(10\cdot5)$, we see that, to the order θ^2 inclusive, we have

$$(11\cdot7) \qquad \begin{cases} 2x\theta^{-2} = \frac{1}{2}(A + C) + \frac{1}{2}(A - C)\cos 2\phi + B\sin 2\phi, \\ 2y\theta^{-2} = \frac{1}{2}(B + D) + \frac{1}{2}(B - D)\cos 2\phi + C\sin 2\phi. \end{cases}$$

Elimination of ϕ from these equations gives an ellipse with centre at the point

(11·8) $\qquad x = \frac{1}{4}(A+C)\theta^2, \quad y = \frac{1}{4}(B+D)\theta^2:$

this is the curve traced out in the focal plane by those rays which make with the principal ray a small angle θ.

If we change θ to θ' in (11·7), but hold ϕ fixed, the corresponding point x', y' is such that

(11·9) $\qquad \dfrac{x'}{x} = \dfrac{y'}{y} = \dfrac{\theta'^2}{\theta^2}.$

Hence, given one of the ellipses (11·7), all the others can be obtained from it by magnification with respect to the origin: the ratio of magnification is positive, and therefore all the points met by rays are to be found on the lines obtained by joining the origin to the points on the ellipse and producing these lines away from the origin. If the origin lies outside the ellipse, these lines are not to be produced through the origin, since this would correspond to a negative ratio of magnification.

Let us investigate the envelope of the ellipses (11·7) for various values of θ. The intersection of consecutive ellipses must satisfy

(11·10) $\qquad \begin{cases} 2xd(\theta^{-2}) = [-(A-C)\sin 2\phi + 2B\cos 2\phi]\,d\phi, \\ 2yd(\theta^{-2}) = [-(B-D)\sin 2\phi + 2C\cos 2\phi]\,d\phi. \end{cases}$

The envelope is therefore to be found by eliminating θ and ϕ from (11·7) and

(11·11) $\qquad \dfrac{x}{y} = \dfrac{(A-C)\sin 2\phi - 2B\cos 2\phi}{(B-D)\sin 2\phi - 2C\cos 2\phi}.$

Comparing this expression with the value of x/y given by (11·7), we see that a real envelope exists if, and only if, ϕ can be found to satisfy

(11·12) $\qquad \dfrac{(A-C)\sin 2\phi - 2B\cos 2\phi}{(B-D)\sin 2\phi - 2C\cos 2\phi}$

$$= \dfrac{A+C+(A-C)\cos 2\phi + 2B\sin 2\phi}{B+D+(B-D)\cos 2\phi + 2C\sin 2\phi}.$$

This equation reduces to

(11·13) $[C(A+C)-B(B+D)]\cos 2\phi + (AD-BC)\sin 2\phi$
$$= B(B-D)-C(A-C),$$

or

(11·14) $\cos(2\phi - \alpha) = \dfrac{B(B-D)-C(A-C)}{[\{C(A+C)-B(B+D)\}^2+(AD-BC)^2]^{\frac{1}{2}}},$

where

(11·15) $\tan \alpha = \dfrac{AD-BC}{C(A+C)-B(B+D)}.$

The condition, necessary and sufficient, for the existence of a real envelope is that the right-hand side of (11·14) should not exceed unity in absolute value: this is expressed analytically by

(11·16) $E \geqslant 0,$

where

(11·17) $E = (AD-BC)^2 - 4(B^2-AC)(C^2-BD).$

Case I. $E > 0$ (Coma).

Since the ellipses are obtained from one another by magnification with respect to the origin (the principal focus), their envelope consists of a pair of lines passing through the focus. The equations

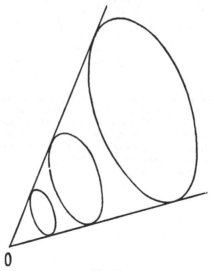

Fig. 22

of the lines may be found by substituting in (11·11) the two values of 2ϕ (in the range 0, 2π) given by (11·14). The ellipses lie only in one of the regions encompassed by the lines, as shown in Fig. 22. Thus the illumination on the focal plane is confined to a wedge-shaped region, the bounding lines being especially bright. This flare of illumination is known as *coma*, from its resemblance to the tail of a comet.

Case II. $E < 0$ (General illumination).

In this case (Fig. 23) the ellipses are contained inside one another and the optical focus is inside them all; the whole region in the neighbourhood of the focus is illuminated.

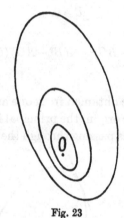

Fig. 23

The criterion distinguishing between coma and general illumination may also be established as follows. Elimination of ϕ from (11·7) gives an equation $F(x, y) = 0$, representing an ellipse, and we have coma or general illumination according as the origin lies outside or inside this ellipse, i.e. according as the product $F(x_0, y_0)\, F(0, 0)$ is negative or positive, x_0 and y_0 being the coordinates of the centre, as in (11·8). Thus a simple algebraic calculation gives $E > 0$ for coma and $E < 0$ for general illumination, with E as in (11·17).

CHAPTER IV

THE INSTRUMENT OF REVOLUTION

12. Approximate form of T for any reflecting or refracting surface of revolution.

Let us take the axis of symmetry of a reflecting or refracting surface of revolution for z-axis. We assume that the equation of the surface may be expanded in the form

$$(12\cdot1) \qquad z = v + \frac{x^2+y^2}{2R} + \frac{(x^2+y^2)^2}{4S} + \dots,$$

where v, R and S are constants, R being the radius of curvature at the vertex $z = v$. We shall confine our attention to rays which are approximately parallel to the axis of symmetry and which meet the surface near the vertex. To the order of approximation which we shall employ it will be unnecessary to consider terms in the expansion (12·1) beyond those shown.

To get a paraboloid of revolution we put

$$(12\cdot2) \qquad\qquad 1/S = 0.$$

If the surface is a sphere of radius R, we have accurately

$$(12\cdot3) \qquad x^2 + y^2 + (z-v-R)^2 = R^2,$$

or, to the above order of approximation,

$$(12\cdot4) \qquad z = v + \frac{x^2+y^2}{2R} + \frac{(x^2+y^2)^2}{8R^3}.$$

Thus, if the surface is a sphere, we are to put

$$(12\cdot5) \qquad\qquad S = 2R^3.$$

Let us now calculate T as a function of the initial components σ', τ' and the final components σ, τ, retaining only terms up to and including the *fourth* order in these components, which are small since the rays are approximately parallel to the axis. By (7·14) we have, accurately,

$$(12\cdot6) \qquad T = (\sigma'-\sigma)\,x + (\tau'-\tau)\,y + (v'-v)\,z,$$

where x, y, z is the point where the ray meets the surface. We are to eliminate x, y, z by the condition that T shall have a stationary value with respect to variations of x, y, z on (12·1).

Let us write

(12·7) $\Delta\sigma = \sigma - \sigma', \quad \Delta\tau = \tau - \tau', \quad \Delta v = v - v',$

Δ indicating in general an increment occurring at the reflection or refraction. Then

(12·8) $T = -x\,\Delta\sigma - y\,\Delta\tau - z\,\Delta v,$

and as in (7·19)

(12·9)
$$
\begin{cases}
\dfrac{\Delta\sigma}{\Delta v} = -\dfrac{\partial z}{\partial x} = -x\left(\dfrac{1}{R} + \dfrac{r^2}{S}\right), \\[2mm]
\dfrac{\Delta\tau}{\Delta v} = -\dfrac{\partial z}{\partial y} = -y\left(\dfrac{1}{R} + \dfrac{r^2}{S}\right),
\end{cases}
$$

$$(r^2 = x^2 + y^2).$$

Thus

(12·10)
$$
\begin{aligned}
x &= -\frac{\Delta\sigma}{\Delta v}\left(\frac{1}{R} + \frac{r^2}{S}\right)^{-1} \\[2mm]
&= -R\frac{\Delta\sigma}{\Delta v}\left(1 - \frac{Rr^2}{S}\right),
\end{aligned}
$$

correct to the third order, and to a first approximation

(12·11) $x = -R\dfrac{\Delta\sigma}{\Delta v}, \quad y = -R\dfrac{\Delta\tau}{\Delta v},$

$$r^2 = R^2\frac{(\Delta\sigma)^2 + (\Delta\tau)^2}{(\Delta v)^2}.$$

Substituting in (12·10), and the similar equation for y, we get, correct to the third order,

(12·12)
$$
\begin{cases}
x = -R\dfrac{\Delta\sigma}{\Delta v}\left(1 - \dfrac{R^3}{S}\dfrac{(\Delta\sigma)^2 + (\Delta\tau)^2}{(\Delta v)^2}\right), \\[3mm]
y = -R\dfrac{\Delta\tau}{\Delta v}\left(1 - \dfrac{R^3}{S}\dfrac{(\Delta\sigma)^2 + (\Delta\tau)^2}{(\Delta v)^2}\right),
\end{cases}
$$

and hence, correct to the fourth order,

(12·13) $r^2 = R^2\dfrac{(\Delta\sigma)^2 + (\Delta\tau)^2}{(\Delta v)^2}\left(1 - \dfrac{2R^3}{S}\dfrac{(\Delta\sigma)^2 + (\Delta\tau)^2}{(\Delta v)^2}\right).$

Then, by (12·1), we have, correct to the fourth order,

$$(12·14) \quad z = v + \frac{r^2}{2R} + \frac{r^4}{4S}$$

$$= v + \tfrac{1}{2}R\frac{(\Delta\sigma)^2 + (\Delta\tau)^2}{(\Delta v)^2}\left(1 - \frac{3}{2}\frac{R^3}{S}\frac{(\Delta\sigma)^2 + (\Delta\tau)^2}{(\Delta v)^2}\right).$$

Substitution in (12·8) from (12·12) and (12·14) gives, correct to the fourth order,

$$(12·15) \quad T = -v\,\Delta v + \tfrac{1}{2}R\frac{(\Delta\sigma)^2 + (\Delta\tau)^2}{\Delta v}\left(1 - \frac{1}{2}\frac{R^3}{S}\frac{(\Delta\sigma)^2 + (\Delta\tau)^2}{(\Delta v)^2}\right).$$

This is the approximate form for T for any reflecting or refracting surface of revolution.

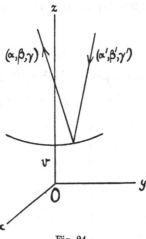

Fig. 24

Let us now consider the case of a *mirror of revolution* with the equation (12·1), the rays being incident from $z = +\infty$ as in Fig. 24, so that the direction cosines of the rays satisfy

$$(12·16) \quad 0 > \gamma' = -(1 - \alpha'^2 - \beta'^2)^{\frac{1}{2}}, \quad 0 < \gamma = (1 - \alpha^2 - \beta^2)^{\frac{1}{2}}.$$

We shall suppose $\mu = 1$. Then

$$(12·17)\quad\begin{cases} \Delta\sigma = \Delta\alpha, \quad \Delta\tau = \Delta\beta, \\ \Delta v = \gamma - \gamma' = 2 - \tfrac{1}{2}(\alpha^2 + \beta^2) - \tfrac{1}{8}(\alpha^2 + \beta^2)^2 \\ \qquad\qquad - \tfrac{1}{2}(\alpha'^2 + \beta'^2) - \tfrac{1}{8}(\alpha'^2 + \beta'^2)^2, \\ \dfrac{1}{\Delta v} = \tfrac{1}{2}[1 + \tfrac{1}{4}(\alpha'^2 + \beta'^2) + \tfrac{1}{4}(\alpha^2 + \beta^2)], \\ \dfrac{1}{(\Delta v)^2} = \tfrac{1}{4}, \end{cases}$$

these expansions being sufficiently accurate for substitution in (12·15). Thus we get, correct to the fourth order,

$$(12·18) \quad T = -2v + \tfrac{1}{2}v(\alpha'^2 + \beta'^2)[1 + \tfrac{1}{4}(\alpha'^2 + \beta'^2)]$$
$$+ \tfrac{1}{2}v(\alpha^2 + \beta^2)[1 + \tfrac{1}{4}(\alpha^2 + \beta^2)] + \tfrac{1}{4}R[(\varDelta\alpha)^2 + (\varDelta\beta)^2]$$
$$\times \left[1 + \tfrac{1}{4}(\alpha'^2 + \beta'^2) + \tfrac{1}{4}(\alpha^2 + \beta^2) - \frac{1}{8}\frac{R^3}{S}\{(\varDelta\alpha)^2 + (\varDelta\beta)^2\}\right].$$

For a *paraboloid* this reduces to

$$(12·19) \quad T = -2v + \tfrac{1}{2}v(\alpha'^2 + \beta'^2)[1 + \tfrac{1}{4}(\alpha'^2 + \beta'^2)]$$
$$+ \tfrac{1}{2}v(\alpha^2 + \beta^2)[1 + \tfrac{1}{4}(\alpha^2 + \beta^2)]$$
$$+ \tfrac{1}{4}R[(\varDelta\alpha)^2 + (\varDelta\beta)^2][1 + \tfrac{1}{4}(\alpha'^2 + \beta'^2) + \tfrac{1}{4}(\alpha^2 + \beta^2)],$$

and for a *sphere*

$$(12·20) \quad T = -2v + \tfrac{1}{2}v(\alpha'^2 + \beta'^2)[1 + \tfrac{1}{4}(\alpha'^2 + \beta'^2)]$$
$$+ \tfrac{1}{2}v(\alpha^2 + \beta^2)[1 + \tfrac{1}{4}(\alpha^2 + \beta^2)] + \tfrac{1}{4}R[(\varDelta\alpha)^2 + (\varDelta\beta)^2]$$
$$\times [1 + \tfrac{1}{4}(\alpha'^2 + \beta'^2) + \tfrac{1}{4}(\alpha^2 + \beta^2) - \tfrac{1}{16}\{(\varDelta\alpha)^2 + (\varDelta\beta)^2\}].$$

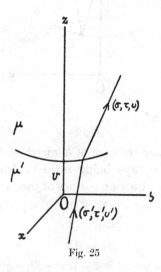

Fig. 25

Let us now consider a *refracting surface of revolution* with the equation (12·1), the rays being incident from $z = -\infty$ as in Fig. 25, so that

$$\begin{cases} 0 < v' = (\mu'^2 - \sigma'^2 - \tau'^2)^{\frac{1}{2}}, \quad 0 < v = (\mu^2 - \sigma^2 - \tau^2)^{\frac{1}{2}}, \\[2mm] \Delta v = v - v' = \mu\left(1 - \dfrac{1}{2}\dfrac{\sigma^2 + \tau^2}{\mu^2} - \dfrac{1}{8}\dfrac{(\sigma^2 + \tau^2)^2}{\mu^4}\right) \\[2mm] \qquad\quad - \mu'\left(1 - \dfrac{1}{2}\dfrac{\sigma'^2 + \tau'^2}{\mu'^2} - \dfrac{1}{8}\dfrac{(\sigma'^2 + \tau'^2)^2}{\mu'^4}\right) \\[2mm] \qquad\quad = \Delta\mu - \tfrac{1}{2}\Delta\dfrac{\sigma^2 + \tau^2}{\mu} - \tfrac{1}{8}\Delta\dfrac{(\sigma^2 + \tau^2)^2}{\mu^3}, \\[2mm] \dfrac{1}{\Delta v} = \dfrac{1}{\Delta\mu} + \dfrac{1}{2}\dfrac{1}{(\Delta\mu)^2}\Delta\dfrac{\sigma^2 + \tau^2}{\mu}, \\[2mm] \dfrac{1}{(\Delta v)^2} = \dfrac{1}{(\Delta\mu)^2}; \end{cases}$$

(12·21)

then (12·15) gives

$$(12\cdot22) \quad T = -v\Delta\mu + \tfrac{1}{2}v\Delta\dfrac{\sigma^2 + \tau^2}{\mu} + \tfrac{1}{8}v\Delta\dfrac{(\sigma^2 + \tau^2)^2}{\mu^3}$$

$$+ \tfrac{1}{2}R\dfrac{(\Delta\sigma)^2 + (\Delta\tau)^2}{\Delta\mu}$$

$$\times \left[1 + \dfrac{1}{2\Delta\mu}\Delta\dfrac{\sigma^2 + \tau^2}{\mu} - \dfrac{1}{2}\dfrac{R^3}{S}\dfrac{(\Delta\sigma)^2 + (\Delta\tau)^2}{(\Delta\mu)^2}\right].$$

For a *sphere* this becomes

$$(12\cdot23) \quad T = -v\Delta\mu + \tfrac{1}{2}v\Delta\dfrac{\sigma^2 + \tau^2}{\mu} + \tfrac{1}{8}v\Delta\dfrac{(\sigma^2 + \tau^2)^2}{\mu^3}$$

$$+ \tfrac{1}{2}R\dfrac{(\Delta\sigma)^2 + (\Delta\tau)^2}{\Delta\mu}$$

$$\times \left[1 + \dfrac{1}{2\Delta\mu}\Delta\dfrac{\sigma^2 + \tau^2}{\mu} - \dfrac{1}{4}\dfrac{(\Delta\sigma)^2 + (\Delta\tau)^2}{(\Delta\mu)^2}\right].$$

13. General form of T: method of calculation up to the fourth order.

An *instrument of revolution* is an instrument with an axis of symmetry, such that the instrument is unchanged (optically) when rotated through any angle about that axis. The most important optical instruments are of this type.

Let us take the axis of symmetry for z-axis: as usual we shall let

$$\mu', \sigma', \tau', \upsilon', \alpha', \beta', \gamma'$$

refer to the initial medium, and

$$\mu, \sigma, \tau, \upsilon, \alpha, \beta, \gamma$$

to the final medium. The T-function for the instrument is a function of the four quantities σ', τ', σ, τ: it may therefore be expressed as a function of the four quantities

(13·1) $\sigma'^2+\tau'^2,\quad \sigma'\sigma+\tau'\tau,\quad \sigma^2+\tau^2,\quad \sigma,$

because these quantities determine σ', τ', σ, τ. The quantities (13·1) in fact determine a ray: if we fix the first three of them and allow σ to vary, we get a single infinity, or pencil, of rays. We have

(13·2) $\begin{cases} \sigma'^2+\tau'^2 = \mu'^2(\alpha'^2+\beta'^2) = \mu'^2(1-\gamma'^2), \\ \sigma'\sigma+\tau'\tau = \mu'\mu(\alpha'\alpha+\beta'\beta) = \mu'\mu(\cos\theta-\gamma'\gamma), \\ \sigma^2+\tau^2 = \mu^2(\alpha^2+\beta^2) = \mu^2(1-\gamma^2), \end{cases}$

where θ is the angle between the initial and final rays. Thus if the first three quantities in (13·1) are given, γ', γ and θ are determined, or, in other words, the inclinations of the initial and final rays to the axis of the instrument and also the mutual inclination of the rays are determined. Now if we take a natural ray passing through the instrument and give it a rigid body rotation about the axis of the instrument, it is clear from the symmetry of the instrument that the ray so obtained will be a natural ray, satisfying the laws of reflection or refraction. But under this rotation γ', γ and θ remain constant: hence the first three quantities in (13·1) remain constant: in fact, the pencil of rays obtained by this rigid body rotation is precisely the pencil obtained by holding these quantities fixed and varying σ. But from its definition as an optical length T is the same for all the rays of the pencil: hence T is actually a function of the first three quantities only in (13·1). Let us put

(13·3) $\epsilon' = \sigma'^2+\tau'^2,\quad \epsilon_, = \sigma'\sigma+\tau'\tau,\quad \epsilon = \sigma^2+\tau^2;$

then

(13·4) $T = T(\epsilon', \epsilon_,, \epsilon).$

The T-function is particularly useful in those cases where the initial rays form a parallel system; when the initial rays diverge from a point source x', y', z' at finite distance it may be more convenient to use the W-function. In general W is a function of

$$x', y', z', \sigma, \tau:$$

by an argument similar to that used above it is easily shown that on account of the symmetry of the instrument W is expressible as a function of the quantities

$$(13\cdot5) \qquad z', \quad x'^2+y'^2, \quad x'\sigma+y'\tau, \quad \sigma^2+\tau^2.$$

We shall use the T-function in what follows.

We shall assume that the rays lie close to the axis of symmetry and are nearly parallel to it. Thus ϵ', $\epsilon_{,}$, ϵ are small, of the second order relative to the inclination of the ray to the axis. We shall suppose that T may be expanded as a power series,

$$(13\cdot6) \qquad T = T^{(0)} + T^{(2)} + T^{(4)} + \ldots,$$

where $T^{(0)}$ is a constant and

$$(13\cdot7) \quad \begin{cases} T^{(2)} = P'\epsilon' + P_{,}\epsilon_{,} + P\epsilon, \\ T^{(4)} = Q''\epsilon'^2 + Q_{,,}\epsilon_{,}^2 + Q\epsilon^2 + Q_{,}'\epsilon'\epsilon_{,} + Q'\epsilon'\epsilon + Q_{,}\epsilon_{,}\epsilon, \end{cases}$$

where the P's and Q's are constants characteristic of the instrument. They depend on the position of the origin on the axis. The superscripts (0), (2), (4) indicate orders of magnitude, the inclination of the ray to the axis of the instrument being the fundamental infinitesimal. In what follows we shall not include terms of order higher than the fourth in T.

As an illustration of the notation employed in (13·7), the results given in (12·18), (12·19), (12·20), (12·22), (12·23) may be exhibited as follows, the surface having the equation

$$(13\cdot8) \qquad z = v + \frac{x^2+y^2}{2R} + \frac{(x^2+y^2)^2}{4S}.$$

For simplicity, the origin is taken at the vertex in some of the formulae given below. To change to a general origin on the axis of the instrument, we may refer to §12, or use (7·30). The directions of the rays are as in Figs. 24 and 25 for reflection and refraction respectively. We note that

$$(13\cdot9) \quad \begin{cases} \textit{in vacuo:} & (\varDelta\alpha)^2+(\varDelta\beta)^2 = \epsilon'-2\epsilon,+\epsilon, \\ \textit{in general:} & (\varDelta\sigma)^2+(\varDelta\tau)^2 = \epsilon'-2\epsilon,+\epsilon. \end{cases}$$

General mirror of revolution with vertex at the origin $(v=0)$.

$$(13\cdot10) \quad \begin{cases} T = \tfrac{1}{4}R(\epsilon'-2\epsilon,+\epsilon)\left[1+\tfrac{1}{4}\epsilon'+\tfrac{1}{4}\epsilon-\dfrac{1}{8}\dfrac{R^3}{S}(\epsilon'-2\epsilon,+\epsilon)\right], \\ P' = P = \tfrac{1}{4}R, \quad P, = -\tfrac{1}{2}R, \\ Q'' = Q = \tfrac{1}{16}R\!\left(1-\dfrac{1}{2}\dfrac{R^3}{S}\right), \quad Q_{\prime\prime} = -\dfrac{1}{8}\dfrac{R^4}{S}, \\ Q', = Q, = \tfrac{1}{8}R\!\left(\dfrac{R^3}{S}-1\right), \quad Q' = \tfrac{1}{8}R\!\left(1-\dfrac{1}{2}\dfrac{R^3}{S}\right). \end{cases}$$

Paraboloidal mirror of revolution with vertex at the origin $(v=0)$.

$$(13\cdot11) \quad \begin{cases} T = \tfrac{1}{4}R(\epsilon'-2\epsilon,+\epsilon)\,(1+\tfrac{1}{4}\epsilon'+\tfrac{1}{4}\epsilon), \\ P' = P = \tfrac{1}{4}R, \quad P, = -\tfrac{1}{2}R, \\ Q'' = Q = \tfrac{1}{16}R, \quad Q_{\prime\prime} = 0, \\ Q', = Q, = -\tfrac{1}{8}R, \quad Q' = \tfrac{1}{8}R. \end{cases}$$

Spherical mirror with vertex at the origin $(v=0)$.

$$(13\cdot12) \quad \begin{cases} T = \tfrac{1}{4}R(\epsilon'-2\epsilon,+\epsilon)\,[1+\tfrac{1}{4}\epsilon'+\tfrac{1}{4}\epsilon-\tfrac{1}{16}(\epsilon'-2\epsilon,+\epsilon)], \\ P' = P = \tfrac{1}{4}R, \quad P, = -\tfrac{1}{2}R, \\ Q'' = Q = \tfrac{3}{64}R, \quad Q_{\prime\prime} = Q', = Q, = -\tfrac{1}{16}R, \quad Q' = \tfrac{3}{32}R. \end{cases}$$

General refracting surface of revolution.

$(13\cdot13)$

$$\begin{cases} T = -v(\mu-\mu')+\tfrac{1}{2}v\mu^{-1}\epsilon-\tfrac{1}{2}v\mu'^{-1}\epsilon'+\tfrac{1}{8}v\mu^{-3}\epsilon^2-\tfrac{1}{8}v\mu'^{-3}\epsilon'^2 \\ \qquad +\dfrac{R}{2(\mu-\mu')}(\epsilon'-2\epsilon,+\epsilon)\left[1+\tfrac{1}{2}(\mu-\mu')^{-1}(\epsilon\mu^{-1}-\epsilon'\mu'^{-1})\right. \\ \qquad\qquad\qquad\qquad\qquad\left. -\dfrac{1}{2}\dfrac{R^3}{S}(\mu-\mu')^{-2}(\epsilon'-2\epsilon,+\epsilon)\right], \\ P' = -\dfrac{1}{2}\dfrac{v}{\mu'}+\dfrac{R}{2(\mu-\mu')}, \quad P = \dfrac{1}{2}\dfrac{v}{\mu}+\dfrac{R}{2(\mu-\mu')}, \quad P, = -\dfrac{R}{\mu-\mu'}, \end{cases}$$

$$Q'' = -\frac{R}{4(\mu-\mu')^2}\left[\mu'^{-1}+\frac{R^3}{S}(\mu-\mu')^{-1}\right]-\frac{1}{8}\frac{v}{\mu'^3},$$

$$Q = \frac{R}{4(\mu-\mu')^2}\left[\mu^{-1}-\frac{R^3}{S}(\mu-\mu')^{-1}\right]+\frac{1}{8}\frac{v}{\mu^3},$$

$$Q_{\prime\prime} = -\frac{R^4}{S(\mu-\mu')^3},$$

$$Q'_{\prime} = \frac{R}{2(\mu-\mu')^2}\left[\mu'^{-1}+\frac{2R^3}{S}(\mu-\mu')^{-1}\right],$$

$$Q_{\prime} = \frac{R}{2(\mu-\mu')^2}\left[-\mu^{-1}+\frac{2R^3}{S}(\mu-\mu')^{-1}\right],$$

$$Q' = -\frac{R}{4(\mu-\mu')}\left[(\mu'\mu)^{-1}+\frac{2R^3}{S}(\mu-\mu')^{-2}\right].$$

Spherical refractor.

(13·14)

$$T = -v(\mu-\mu')+\tfrac{1}{2}v\mu^{-1}\epsilon-\tfrac{1}{2}v\mu'^{-1}\epsilon'+\tfrac{1}{8}v\mu^{-3}\epsilon^2-\tfrac{1}{8}v\mu'^{-3}\epsilon'^2$$

$$+\frac{R}{2(\mu-\mu')}(\epsilon'-2\epsilon_{\prime}+\epsilon)\,[1+\tfrac{1}{2}(\mu-\mu')^{-1}(\epsilon\mu^{-1}-\epsilon'\mu'^{-1})$$

$$-\tfrac{1}{4}(\mu-\mu')^{-2}(\epsilon'-2\epsilon_{\prime}+\epsilon)],$$

$$P' = -\frac{1}{2}\frac{v}{\mu'}+\frac{R}{2(\mu-\mu')},\quad P = \frac{1}{2}\frac{v}{\mu}+\frac{R}{2(\mu-\mu')},\quad P_{\prime} = -\frac{R}{\mu-\mu'},$$

$$Q'' = -\frac{R}{4(\mu-\mu')^2}[\mu'^{-1}+\tfrac{1}{2}(\mu-\mu')^{-1}]-\frac{1}{8}\frac{v}{\mu'^3}$$

$$= -\frac{R(2\mu-\mu')}{8\mu'(\mu-\mu')^3}-\frac{1}{8}\frac{v}{\mu'^3},$$

$$Q = \frac{R}{4(\mu-\mu')^2}[\mu^{-1}-\tfrac{1}{2}(\mu-\mu')^{-1}]+\frac{1}{8}\frac{v}{\mu^3} = \frac{R(\mu-2\mu')}{8\mu(\mu-\mu')^3}+\frac{1}{8}\frac{v}{\mu^3},$$

$$Q_{\prime\prime} = -\frac{1}{2}\frac{R}{(\mu-\mu')^3},$$

$$Q'_{\prime} = \frac{R}{2(\mu-\mu')^2}[\mu'^{-1}+(\mu-\mu')^{-1}] = \frac{R\mu}{2\mu'(\mu-\mu')^3},$$

$$Q_{\prime} = \frac{R}{2(\mu-\mu')^2}[-\mu^{-1}+(\mu-\mu')^{-1}] = \frac{R\mu'}{2\mu(\mu-\mu')^3},$$

$$Q' = -\frac{R}{4(\mu-\mu')}[(\mu\mu')^{-1}+(\mu-\mu')^{-2}] = -\frac{R(\mu^2-\mu\mu'+\mu'^2)}{4\mu\mu'(\mu-\mu')^3}.$$

In § 7 we had a general method for the calculation of the T-function for any instrument. Two steps were involved: (i) the determination of the T-function for each pair of successive media, (ii) the elimination of the intermediate components, so as to leave T a function of the initial and final components. Now we are only interested in the calculation of T up to the fourth order for an instrument of revolution. The general process may be simplified under these circumstances.

In the notation of (7·28) the T-function for the complete instrument is

(13·15) $$T = T_{0,1} + T_{1,2} + \ldots + T_{n-1,n};$$

the intermediate components are to be eliminated by means of

(13·16) $$\frac{\partial T}{\partial \sigma_i} = 0, \quad \frac{\partial T}{\partial \tau_i} = 0, \qquad (i = 1, 2, \ldots, n-1).$$

Now T as given in (13·15) may be expanded in the form

(13·17) $$T = T^{(0)} + T^{(2)} + T^{(4)} + \ldots,$$

the superscripts indicating orders of magnitude in the small components. If we retain only terms up to the fourth order in T, the equations (13·16) are of the third degree: we shall see, however, that it is sufficient to eliminate the intermediate variables by means of linear equations, in accordance with the following theorem: *If instead of using the exact equations* (13·16) *we eliminate the intermediate components from* (13·15) *by means of the linear equations*

(13·18) $$\frac{\partial T^{(2)}}{\partial \sigma_i} = 0, \quad \frac{\partial T^{(2)}}{\partial \tau_i} = 0, \quad (i = 1, 2, \ldots, n-1),$$

the error so introduced in T is of the sixth order, and is therefore negligible when T is required only to the fourth order.

It is simplest to prove this theorem in a more general form. Let $x_1, x_2, \ldots, x_m$ be a set of small variables (corresponding to the intermediate components) and $y_1, y_2, \ldots, y_n$ another set of small variables (corresponding to the initial and final components). Let f (corresponding to T) be a function of the x's and y's, and let it be represented by a series of the form

(13·19) $$f(x, y) = f^{(0)} + f^{(2)}(x, y) + f^{(4)}(x, y) + \ldots,$$

$f^{(0)}$ being a constant and $f^{(2)}$, $f^{(4)}$ polynomials, homogeneous of degrees 2 and 4 respectively. We are to eliminate the x's from f by two different processes. Let us in general write

$$(13\cdot20) \qquad \phi_r(x,y) = \frac{\partial \phi(x,y)}{\partial x_r}.$$

The first process of elimination is by means of

$$(13\cdot21) \qquad f_r^{(2)}(x,y) = 0 \quad (r = 1, 2, ..., m).$$

Let the solution of these equations be

$$(13\cdot22) \qquad x_r = \xi_r(y);$$

when these are substituted in $f(x,y)$, we get

$$(13\cdot23) \qquad F(y) = f(\xi, y).$$

We have of course

$$(13\cdot24) \qquad f_r^{(2)}(\xi, y) \equiv 0,$$

for arbitrary values of the y's.

On the other hand, let us eliminate the x's by means of

$$(13\cdot25) \qquad f_r(x,y) = 0.$$

Let the solution of these equations be

$$(13\cdot26) \qquad x_r = \xi_r(y) + \eta_r(y);$$

when these are substituted in $f(x,y)$ we get

$$(13\cdot27) \qquad G(y) = f(\xi + \eta, y).$$

We have of course

$$(13\cdot28) \qquad f_r(\xi + \eta, y) \equiv 0,$$

for arbitrary values of the y's. *We shall now prove that*

$$G(y) - F(y)$$

is of the sixth order.

The equation (13·28) may be written

$$(13\cdot29) \qquad f_r^{(2)}(\xi + \eta, y) + f_r^{(4)}(\xi + \eta, y) + ... = 0,$$

or, expanding the first term as a power series in the η's,

$$(13\cdot30) \quad f_r^{(2)}(\xi, y) + \sum_{s=1}^{m} \eta_s f_{rs}^{(2)}(\xi, y) + ... + f_r^{(4)}(\xi + \eta, y) + ... = 0,$$

where $f_{rs}^{(2)}$ is a partial derivative of the second order. The first term vanishes, by (13·24): $f_{rs}^{(2)}$ are constants: the term $f_r^{(4)}$ is of the third order. Hence the η's are of the third order. We have then, writing O_6 for terms of the sixth order,

$$(13\cdot31) \quad G(y) - F(y) = f(\xi + \eta, y) - f(\xi, y)$$
$$= f^{(2)}(\xi + \eta, y) - f^{(2)}(\xi, y)$$
$$+ f^{(4)}(\xi + \eta, y) - f^{(4)}(\xi, y) + O_6$$
$$= \tfrac{1}{2} \sum_{r=1}^{m} \sum_{s=1}^{m} \eta_r \eta_s f_{rs}^{(2)}(\xi, y) + \sum_{r=1}^{m} \eta_r f_r^{(4)}(\xi, y) + O_6$$
$$= O_6;$$

this establishes the result, and hence proves the theorem associated with (13·18).

The simplification consequent on the use of linear equations (13·18) instead of cubic equations is naturally very great.

14. First order theory: object and image points: cardinal points.

We shall now develop the first order theory of a general instrument of revolution, the rays being adjacent to the axis of the instrument. In this approximation we shall neglect in the equations of the rays the squares and higher powers of the distances from the axis of the instrument and of the inclinations of the rays to the axis. It is therefore only necessary to retain terms of the second order in T.

Taking the z-axis along the axis of the instrument, we have to the required order of approximation

$$(14\cdot1) \qquad T = T^{(0)} + P'\epsilon' + P_\prime \epsilon_\prime + P\epsilon + \text{const.},$$

$$\epsilon' = \sigma'^2 + \tau'^2, \quad \epsilon_\prime = \sigma'\sigma + \tau'\tau, \quad \epsilon = \sigma^2 + \tau^2,$$

as in (13·7). To the order of approximation here considered, the three constants P', $P_\prime$, P and the initial and final refractive indices determine the optical behaviour of the instrument. We shall suppose these constants known, and develop the optical properties in terms of them.

The equations of the initial and final rays are given by (7·8), (7·9), in which x', y', x, y, σ', τ', σ, τ are small. Since

$$(14\cdot2) \qquad v'^2 = \mu'^2 - \sigma'^2 - \tau'^2, \quad v^2 = \mu^2 - \sigma^2 - \tau^2,$$

we have to the first order

(14·3) $$v' = \eta'\mu', \quad v = \eta\mu,$$

where η', η are ± 1, being $+1$ if the ray in question is proceeding in the positive sense of the z-axis and -1 if it is proceeding in the negative sense. When only refractions occur, η' and η have the same sign, so that $\eta'\eta = 1$. More generally, this condition holds when the final rays have the same sense as the initial rays. We may then say that the instrument is *direct*. When the instrument is a single mirror, we have $\eta'\eta = -1$. More generally this condition holds when the final rays have a sense opposite to that of the initial rays. We may then say that the instrument is *reversing*. By retaining the factors η', η we shall be able to discuss both types of instrument at once.

By (7·8), then, the equations of the initial rays are approximately

(14·4)
$$\begin{cases} x' - \eta'z'\sigma'/\mu' = \dfrac{\partial T}{\partial \sigma'} = 2P'\sigma' + P_{,}\sigma, \\[2ex] y' - \eta'z'\tau'/\mu' = \dfrac{\partial T}{\partial \tau'} = 2P'\tau' + P_{,}\tau, \end{cases}$$

and the equations of the final rays are, by (7·9), approximately

(14·5)
$$\begin{cases} x - \eta z\sigma/\mu = -\dfrac{\partial T}{\partial \sigma} = -P_{,}\sigma' - 2P\sigma, \\[2ex] y - \eta z\tau/\mu = -\dfrac{\partial T}{\partial \tau} = -P_{,}\tau' - 2P\tau. \end{cases}$$

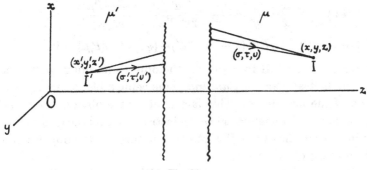

Fig. 26

Let us suppose that there is a point source of light at a point I' near the axis of the instrument, the coordinates of I' being x', y', z' (Fig. 26). In (14·4), (14·5) we have four equations: elimination of σ', τ' will give the equations of the final congruence of rays due to the source I'. By (14·4) we have

$$(14·6) \qquad \sigma' = \frac{x' - P_, \sigma}{2P' + \eta'z'/\mu'}, \quad \tau' = \frac{y' - P_, \tau}{2P' + \eta'z'/\mu'}.$$

Substituting in (14·5) and rearranging, we have

$$(14·7) \quad \begin{cases} x + \dfrac{P_, x'}{2P' + \eta'z'/\mu'} = \sigma \left(\eta z/\mu - 2P + \dfrac{P_,^2}{2P' + \eta'z'/\mu'} \right), \\[3mm] y + \dfrac{P_, y'}{2P' + \eta'z'/\mu'} = \tau \left(\eta z/\mu - 2P + \dfrac{P_,^2}{2P' + \eta'z'/\mu'} \right). \end{cases}$$

No matter what values σ and τ may have, these equations are satisfied by

$$(14·8) \quad \begin{cases} x = -\dfrac{P_, x'}{2P' + \eta'z'/\mu'}, \quad y = -\dfrac{P_, y'}{2P' + \eta'z'/\mu'}, \\[3mm] z = \eta\mu \left(2P - \dfrac{P_,^2}{2P' + \eta'z'/\mu'} \right). \end{cases}$$

Thus, to the first order, all the rays emanating from an object point $I'(x', y', z')$ in the initial medium pass (after traversing the instrument) through an image point $I(x, y, z)$, with coordinates given by (14·8).

The equations (14·8) may be written more symmetrically

$$(14·9) \qquad \frac{x}{x'} = \frac{y}{y'} = -\frac{\eta'P_,\mu'}{z' + 2\eta'P'\mu'} = \frac{z - 2\eta P\mu}{\eta P_,\mu},$$

$$(14·10) \quad (z - 2\eta P\mu)(z' + 2\eta'P'\mu') + (\eta'P_,\mu')(\eta P_,\mu) = 0.$$

Since $x/y = x'/y'$, it follows that the object and image points lie in a diametral plane, a *diametral plane* being a plane through the axis of the instrument. This fact is of course obvious from the symmetry of the system, since the instrument and the congruence of initial rays have the diametral plane through the object point for a plane of symmetry.

It is also clear from symmetry that if the object point lies on the axis of the instrument, so also does the image point: this also follows from (14·8), since $x' = y' = 0$ imply $x = y = 0$. Points on the axis such that one is the image of the other are called *conjugate points*: the relation between the positions of conjugate points is (14·10). The planes through conjugate points perpendicular to the axis of the instrument are called *conjugate planes*. It is evident that an object point and its image are situated in conjugate planes.

We shall now define three pairs of *cardinal points* on the axis of the instrument. These are the *focal points*, the *nodal points* and the *principal points*, the planes through them perpendicular to the axis of the instrument being the *focal*, *nodal* and *principal planes*.

The *focal point* F' is defined as the object point whose image point is at an infinite distance on the axis of the instrument, and the *focal point* F is defined as the image point whose object point is at an infinite distance on the axis of the instrument. Thus the z' of F' is to be found by letting $z \to \infty$ in (14·10) and the z of F by letting $z' \to \infty$ in the same equation. Hence we have for the focal points

$$(14\cdot11) \qquad z(F') = -2\eta'P'\mu', \quad z(F) = 2\eta P\mu.$$

Since object and image points are principal foci in the sense of § 11, and in particular so are focal points, we may say that

the focal point F' is the principal focus in the initial medium for rays parallel to the axis in the final medium;

the focal point F is the principal focus in the final medium for rays parallel to the axis in the initial medium.

The *nodal points* N', N are defined as conjugate points such that the corresponding rays through them are parallel (in the same or opposite senses). If the senses are the same, we are to have $\alpha' = \alpha$, $\beta' = \beta$, and if they are opposite, $\alpha' = -\alpha$, $\beta' = -\beta$. Thus the conditions for the two cases are included in

$$(14\cdot12) \qquad \eta'\alpha' = \eta\alpha, \quad \eta'\beta' = \eta\beta,$$

α', β' and α, β being direction cosines of corresponding rays through N' and N respectively. Substituting in (14·4), (14·5) the following values,

$$(14·13) \quad \begin{cases} x' = y' = x = y = 0, \\ \sigma' = \mu'\alpha', \quad \tau' = \mu'\beta', \quad \sigma = \mu\alpha, \quad \tau = \mu\beta, \end{cases}$$

we have

$$(14·14) \quad \begin{cases} -\eta'z'\alpha' = 2P'\mu'\alpha' + P_{,}\mu\alpha, \\ -\eta z\alpha = -P_{,}\mu'\alpha' - 2P\mu\alpha, \end{cases}$$

with similar equations with β', β substituted for α', α. Hence by (14·12) we obtain for the positions of the nodal points

$$(14·15) \quad \begin{cases} z(N') = -2\eta'P'\mu' - \eta P_{,}\mu, \\ z(N) = 2\eta P\mu + \eta'P_{,}\mu'. \end{cases}$$

Let us now take as object a short line $A'B'$, perpendicular to the axis, A' being on the axis. Let A be the image of A' and B the

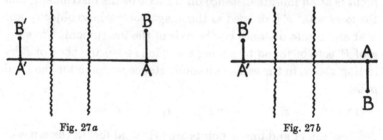

Fig. 27a   Fig. 27b

image of B'. We know by (14·8) that AB will be parallel to $A'B'$: all the points on $A'B'$ will have images on AB, and we may speak of the line AB as the *image* of the line $A'B'$. The line AB may have the same sense as $A'B'$ (Fig. 27a) or the opposite sense (Fig. 27b). In the former cases we have an *erect* image, in the latter an *inverted* image.

We define the *magnification* (m) to be $\pm AB/A'B'$, the $+$ sign being taken when the image is erect, the $-$ when inverted.† By (14·9) the magnification is given by

$$(14·16) \quad m = \frac{x}{x'} = \frac{y}{y'} = -\frac{\eta'P_{,}\mu'}{z' + 2\eta'P'\mu'} = \frac{z - 2\eta P\mu}{\eta P_{,}\mu}.$$

The *principal points* U', U are defined to be conjugate points

† Or we may say $m = AB/A'B'$, interpreted algebraically.

for which the magnification is equal to unity. Hence, putting $m = 1$ in (14·16), we have

$$(14·17) \quad \begin{cases} z(U') = -\eta'\mu'(2P'+P_{,}), \\ z(U) = \eta\mu(2P+P_{,}). \end{cases}$$

By (14·5) we have, for any ray through U,

$$(14·18) \quad \begin{cases} -\sigma(2P+P_{,}) = -P_{,}\sigma' - 2P\sigma, \\ -\tau(2P+P_{,}) = -P_{,}\tau' - 2P\tau. \end{cases}$$

Hence *for corresponding rays through the principal points the relations* $\sigma' = \sigma$, $\tau' = \tau$ *hold.* These conditions might have been used as a definition of principal points.

Let us put for brevity

$$(14·19) \quad \begin{cases} 2\eta'\mu'P' = a', \quad \eta'\mu'P_{,} = b', \\ 2\eta\mu P = a, \qquad \eta\mu P_{,} = b. \end{cases}$$

Then by (14·11), (14·15), (14·17) the positions of the cardinal points are as follows:

$$(14·20) \quad \begin{cases} z(F') = -a', \qquad z(F) = a, \\ z(N') = -a'-b, \quad z(N) = a+b', \\ z(U') = -a'-b', \quad z(U) = a+b. \end{cases}$$

Hence

$$(14·21) \quad \begin{cases} z(N') - z(U') = z(N) - z(U), \\ z(N') - z(F') = z(F) - z(U), \\ z(N) - z(F) = z(F') - z(U'). \end{cases}$$

Using the ordinary notation for directed segments on a line, in which a segment is counted positive or negative according as it runs in the positive or negative sense, we have

$$(14·22) \quad U'N' = UN, \quad F'N' = UF, \quad FN = U'F'.$$

A possible arrangement of the cardinal points is shown in Fig. 28.

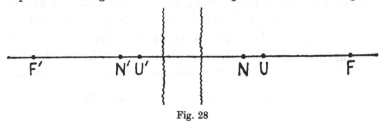

Fig. 28

The following relations are also easily proved:

(14·23) $\eta\mu FN + \eta'\mu'F'N' = 0$, $\eta\mu F'U' + \eta'\mu'FU = 0$.

Further, if C', C is any pair of conjugate points, it follows from (14·10) that

(14·24) $FC \cdot F'C' = FN \cdot F'N' = FU \cdot F'U'$.

The pairs of conjugate points form two homographic ranges on the axis of the instrument,† the relation (14·10) being, in the notation of (14·19),

(14·25) $(z-a)(z'+a') + bb' = 0$.

The double points, i.e. those points which are their own conjugates, are found by putting $z' = z$: thus

(14·26) $z^2 + z(a'-a) - aa' + bb' = 0$,

$$z = \tfrac{1}{2}\{z(F') + z(F)\} \pm \sqrt{\tfrac{1}{4}F'F^2 + F'N' \cdot FN}.$$

Such points necessarily exist if $bb' < 0$, which is the case if η', η have opposite signs, i.e. if the instrument is of the reversing type.

The first and second *focal lengths* of the instrument are defined as

(14·27) $f' = F'U'$, $f = UF$.

Thus, by (14·20),

(14·28) $f' = -\eta'\mu'P_{,}$, $f = -\eta\mu P_{,}$,

and so the focal lengths are connected by the equation

(14·29) $\eta'f'/\mu' = \eta f/\mu$.

We observe that when the initial and final refractive indices are the same ($\mu' = \mu$) and the instrument is of the direct type ($\eta' = \eta$), we have $f' = f$. Further, in this case the nodal points coincide with the principal points.

The image of a given object can be constructed very simply when the focal and principal points are given. Let $A'B'$ be the object (Fig. 29). Through B' draw a parallel to the axis of the instrument, cutting the principal planes at V', V respectively. The incident ray $B'V'$ emerges as VF. Through B' draw $B'F'$, cutting the principal plane through U' at W'. Through W' draw a line parallel to the axis of the instrument, cutting the principal

† Cf. C. V. Durell, *Plane Geometry*, Part II (London, 1910), p. 206.

plane through U at W. The incident ray $B'F'$ emerges along the line just drawn, and the image of B' is the intersection B of this line and VF.

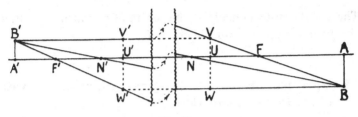

Fig. 29

If the nodal points N', N are given we may use a ray through them instead of one of the two rays given above, remembering that the final part of the ray through N is parallel to the initial part through N'. Thus we may carry out our construction using any one of the following sets of points: $F'FU'U$, $F'U'N'N$, $FUN'N$.

As a simple illustration of some of the preceding formulae, we may apply them to the case of a mirror of radius of curvature R with approximate equation

$$(14 \cdot 30) \qquad z = \frac{x^2 + y^2}{2R}.$$

We have by (13·12)

$$(14 \cdot 31) \qquad P' = P = \tfrac{1}{4}R, \quad P_{,} = -\tfrac{1}{2}R,$$

and $\mu' = \mu = 1$, $\eta' = \!\!\!-1$, $\eta = 1$, the incident rays travelling in the negative sense as in Fig. 24, in which now $v = 0$. We find from formulae given above that the focal points coincide at $z = \tfrac{1}{2}R$, the nodal points coincide at $z = R$ (the centre of curvature), and the principal points coincide at $z = 0$ (on the mirror).

Taking (14·30) for the surface of separation of media of refractive indices μ' for $z < 0$ and μ for $z > 0$, as in Fig. 25 with $v = 0$, we have as in (13·13) for $v = 0$

$$(14 \cdot 32) \qquad P' = P = \frac{R}{2(\mu - \mu')}, \quad P_{,} = -\frac{R}{\mu - \mu'}.$$

We are also to put $\eta' = \eta = 1$. We find for the focal points

$$(14\cdot33) \qquad z(F') = -\frac{\mu'}{\mu - \mu'}\, R, \quad z(F) = \frac{\mu}{\mu - \mu'}\, R.$$

The nodal points coincide at the centre of curvature $z = R$, and the principal points coincide at $z = 0$.

15. Spherical aberration, astigmatism, coma, curvature of the image, distortion.

We have seen that if, for rays adjacent to the axis of an instrument of revolution, we neglect in the equations of the rays small quantities of order higher than the first, then to each object point there corresponds an image point. In fact, if the incident rays emanate from a point, the final rays pass (to the first order) through a point.

One of the purposes of an optical instrument is to produce a point image of a point object. To the first order any instrument of revolution does this for monochromatic light. But when terms of higher orders are taken into consideration this is no longer the case. Furthermore, given a small object-pattern on a plane perpendicular to the axis, a perfect instrument should produce on some plane an image-pattern, in which the dimensions of the object-pattern are magnified or diminished uniformly. It is evident from (14·9) that, to the first order, every instrument of revolution is perfect in this respect. But, again, defects appear when a more accurate discussion is given.

Confining our attention to monochromatic light or to an instrument involving reflections only, so that chromatic aberrations (§ 19) do not occur, we have to consider *five* defects in an instrument of revolution. The first three of these, *spherical aberration*, *astigmatism* and *coma*, arise from failure to produce a point image The other two, *curvature of the image* and *distortion*, concern the failure of the instrument to reproduce a plane pattern to scale.

In the approximation now to be given we shall include terms of the fourth order in T, but omit terms of the sixth order. Thus

$$(15 \cdot 1) \quad T = T^{(0)} + P'\epsilon' + P_{,}\epsilon_{,} + P\epsilon + Q''\epsilon'^2 + Q_{,,}\epsilon_{,}^2 + Q\epsilon^2$$
$$+ Q'_{,}\epsilon'\epsilon_{,} + Q'\epsilon'\epsilon + Q_{,}\epsilon_{,}\epsilon,$$
$$\epsilon' = \sigma'^2 + \tau'^2, \quad \epsilon_{,} = \sigma'\sigma + \tau'\tau, \quad \epsilon = \sigma^2 + \tau^2.$$

We shall limit our attention to *object points at an infinite distance*, so that the congruence of incident rays from an object point consists of parallel rays; σ' and τ' are then the same for all these rays. To deal with objects at finite distance it is more convenient to use the characteristic function W.

We shall assume that the final rays travel in the positive sense, so that $v > 0$. We have then approximately

$$(15 \cdot 2) \qquad \frac{1}{v} = \frac{1}{(\mu^2 - \epsilon)^{\frac{1}{2}}} = \mu^{-1}(1 + \tfrac{1}{2}\epsilon\mu^{-2}).$$

Since

$$(15 \cdot 3)$$

$$\frac{\partial \epsilon'}{\partial \sigma} = \frac{\partial \epsilon'}{\partial \tau} = 0, \quad \frac{\partial \epsilon_{,}}{\partial \sigma} = \sigma', \quad \frac{\partial \epsilon_{,}}{\partial \tau} = \tau', \quad \frac{\partial \epsilon}{\partial \sigma} = 2\sigma, \quad \frac{\partial \epsilon}{\partial \tau} = 2\tau,$$

the exact equations (7·9) of the final rays, namely,

$$(15 \cdot 4) \qquad -x + z\frac{\sigma}{v} = \frac{\partial T}{\partial \sigma}, \quad -y + z\frac{\tau}{v} = \frac{\partial T}{\partial \tau},$$

give, to the third order inclusive,

$$(15 \cdot 5) \quad \begin{cases} -x + z\sigma\mu^{-1}(1 + \tfrac{1}{2}\epsilon\mu^{-2}) = P_{,}\sigma' + 2P\sigma + 2Q_{,,}\epsilon_{,}\sigma' \\ \qquad + 4Q\epsilon\sigma + Q'_{,}\epsilon'\sigma' + 2Q'\epsilon'\sigma + Q_{,}(\sigma'\epsilon + 2\epsilon_{,}\sigma), \\ -y + z\tau\mu^{-1}(1 + \tfrac{1}{2}\epsilon\mu^{-2}) = P_{,}\tau' + 2P\tau + 2Q_{,,}\epsilon_{,}\tau' \\ \qquad + 4Q\epsilon\tau + Q'_{,}\epsilon'\tau' + 2Q'\epsilon'\tau + Q_{,}(\tau'\epsilon + 2\epsilon_{,}\tau). \end{cases}$$

The terms omitted are of the fifth order at least.

Spherical aberration.

Let there be an object point at an infinite distance on the axis of the instrument. The incident rays are then parallel to the axis, and we have

$$(15 \cdot 6) \qquad \sigma' = \tau' = 0, \quad \epsilon' = \epsilon_{,} = 0.$$

Hence by (15·5) the final rays have the equations

$$(15 \cdot 7) \quad \begin{cases} -x + z\sigma\mu^{-1}(1 + \tfrac{1}{2}\epsilon\mu^{-2}) = 2P\sigma + 4Q\epsilon\sigma, \\ -y + z\tau\mu^{-1}(1 + \tfrac{1}{2}\epsilon\mu^{-2}) = 2P\tau + 4Q\epsilon\tau. \end{cases}$$

The final ray with components σ, τ cuts the plane $z = $ const. in the point

$$(15\cdot8) \quad \begin{cases} x = \sigma\mu^{-1}(z - 2P\mu) + \tfrac{1}{2}\sigma\epsilon\mu^{-3}(z - 8Q\mu^3), \\ y = \tau\mu^{-1}(z - 2P\mu) + \tfrac{1}{2}\tau\epsilon\mu^{-3}(z - 8Q\mu^3). \end{cases}$$

In general this point has a first-order distance from the axis, but if the plane $z = $ const. is the focal plane $z = 2P\mu$ (14·11), this distance is of the third order: we have then

$$(15\cdot9) \quad \begin{cases} x = \sigma\epsilon\mu^{-2}(P - 4Q\mu^2), \\ y = \tau\epsilon\mu^{-2}(P - 4Q\mu^2), \end{cases}$$

the distance from the axis being

$$(15\cdot10) \qquad r = \mid \epsilon^{\frac{1}{2}}\mu^{-2}(P - 4Q\mu^2) \mid.$$

The rays (15·7) for arbitrary σ, τ do not pass through a point. There is a principal focus or first-order image at $x = y = 0$, $z = 2P\mu$, but final rays inclined to the axis of the instrument pass by this point at a distance (15·10) which is of the third order, and so not negligible for the present approximation. This deviation from the point image for incident rays parallel to the axis of the instrument is called *spherical aberration*. It is obvious from (15·9) that it is present unless the instrument is designed to satisfy the condition

$$(15\cdot11) \qquad P = 4Q\mu^2.$$

This is the condition for the absence of spherical aberration.

Let us consider a *mirror of revolution*, for which we have, as in (13·10),

$$(15\cdot12) \quad \mu = 1, \quad P = \tfrac{1}{4}R, \quad Q = \tfrac{1}{16}R\left(1 - \frac{1}{2}\frac{R^3}{S}\right).$$

Then

$$(15\cdot13) \qquad\qquad P - 4Q\mu^2 = \frac{1}{8}\frac{R^4}{S}.$$

The equation of the mirror is (13·8), with $v = 0$. The defect of spherical aberration will therefore be present in a mirror unless $S = \infty$. This condition gives a paraboloidal mirror, for which indeed we know from elementary geometrical considerations or from (8·18) that rays parallel to the axis are reflected accurately through the geometrical focus.

In the case of a *spherical mirror* of radius R, we have as in (12·5) $S = 2R^3$, and hence

(15·14) $$P - 4Q\mu^2 = \tfrac{1}{16}R.$$

It is evident from (15·9) that for any instrument of revolution a final ray inclined to the axis of the instrument at an angle θ cuts that axis at a distance

$$\mu\theta^2(P - 4Q\mu^2)$$

in front of the focal point, i.e. at

(15·15) $$z = 2P\mu - \mu\theta^2(P - 4Q\mu^2);$$

this also follows from (15·8) on putting $x = 0$. For the case of a spherical mirror this result is easily checked by trigonometry.

Conditions for the formation of a point image.

We have already seen that the condition (15·11) is necessary and sufficient for the formation of a point image of an object point at infinite distance on the axis, to the order of approximation considered. We shall now show that *if* (15·11) *is satisfied and also*

(15·16) $$Q_{,} = Q_{,,} = 0,$$

then any congruence of parallel incident rays at small inclination to the axis gives a point image. In this case the object point is at infinity, but not on the axis.

Substituting from (15·16) in (15·5), we obtain

(15·17)
$$\begin{cases} -x + z\sigma\mu^{-1}(1 + \tfrac{1}{2}\epsilon\mu^{-2}) = P_{,}\sigma' + 2P\sigma \\ \qquad\qquad\qquad\qquad + 4Q\epsilon\sigma + Q'_{,}\epsilon'\sigma' + 2Q'\epsilon'\sigma, \\ -y + z\tau\mu^{-1}(1 + \tfrac{1}{2}\epsilon\mu^{-2}) = P_{,}\tau' + 2P\tau \\ \qquad\qquad\qquad\qquad + 4Q\epsilon\tau + Q'_{,}\epsilon'\tau' + 2Q'\epsilon'\tau, \end{cases}$$

as the equations of the final rays. The first of these may be written, by (15·11),

(15·18) $$x + P_{,}\sigma' + Q'_{,}\epsilon'\sigma' = \sigma\{(1 + \tfrac{1}{2}\epsilon\mu^{-2})(z\mu^{-1} - 2P) - 2Q'\epsilon'\}.$$

We are justified, to the order of approximation employed, in adding a term of the fifth order, since this is negligible: hence (15·18) and its companion may be written

(15·19)
$$\begin{cases} x + P_{,}\sigma' + Q'_{,}\epsilon'\sigma' = \sigma(1 + \tfrac{1}{2}\epsilon\mu^{-2})(z\mu^{-1} - 2P - 2Q'\epsilon'), \\ y + P_{,}\tau' + Q'_{,}\epsilon'\tau' = \tau(1 + \tfrac{1}{2}\epsilon\mu^{-2})(z\mu^{-1} - 2P - 2Q'\epsilon'). \end{cases}$$

To the order of approximation considered, all these final rays pass through the point

$$(15 \cdot 20) \quad \begin{cases} x = -P_{,}\sigma' - Q'_{,}\epsilon'\sigma', \\ y = -P_{,}\tau' - Q'_{,}\epsilon'\tau', \\ z = 2P\mu + 2Q'\epsilon'\mu. \end{cases}$$

This establishes the result stated.

By treating $(15 \cdot 5)$ as identities in σ, τ, it is easily seen that the conditions $(15 \cdot 11)$, $(15 \cdot 16)$ are necessary as well as sufficient for the formation of a point image.

Astigmatism.

Let us now suppose that

$$(15 \cdot 21) \quad P = 4Q\mu^2, \quad Q_{,} = 0, \quad Q_{,,} \neq 0.$$

Employing the same device as before, namely, the addition of terms of the fifth order, the equations $(15 \cdot 5)$ for the final rays may be written

$$(15 \cdot 22) \quad \begin{cases} x + P_{,}\sigma' + Q'_{,}\epsilon'\sigma' - \sigma(1 + \tfrac{1}{2}\epsilon\mu^{-2})(z\mu^{-1} - 2P - 2Q'\epsilon') \\ \qquad = -2Q_{,,}\epsilon_{,}\sigma', \\ y + P_{,}\tau' + Q'_{,}\epsilon'\tau' - \tau(1 + \tfrac{1}{2}\epsilon\mu^{-2})(z\mu^{-1} - 2P - 2Q'\epsilon') \\ \qquad = -2Q_{,,}\epsilon_{,}\tau'. \end{cases}$$

Let us choose the axes of x and y so that the plane of xz is parallel to the incident rays: then

$$(15 \cdot 23) \quad \tau' = 0, \quad \epsilon_{,} = \sigma\sigma', \quad \epsilon' = \sigma'^2,$$

and $(15 \cdot 22)$ become

$$(15 \cdot 24) \quad \begin{cases} x + P_{,}\sigma' + Q'_{,}\sigma'^3 \\ \qquad = \sigma(1 + \tfrac{1}{2}\epsilon\mu^{-2})(z\mu^{-1} - 2P - 2Q'\sigma'^2 - 2Q_{,,}\sigma'^2), \\ y = \tau(1 + \tfrac{1}{2}\epsilon\mu^{-2})(z\mu^{-1} - 2P - 2Q'\sigma'^2). \end{cases}$$

Consider the lines

$$(15 \cdot 25) \quad \begin{cases} x + P_{,}\sigma' + Q'_{,}\sigma'^3 = 0, \\ z - 2\mu(P + Q'\sigma'^2 + Q_{,,}\sigma'^2) = 0, \end{cases} \Bigg\} L_1;$$

$$\begin{cases} y = 0, \\ z - 2\mu(P + Q'\sigma'^2) = 0, \end{cases} \Bigg\} L_2.$$

No matter what values σ and τ may have, we can find x, z to satisfy the equations of L_1, and therefore the first of (15·24); we can then choose y to satisfy the second of (15·24). Thus the ray (15·24) cuts L_1. Similarly, it cuts L_2. *Thus all the final rays cut the two lines L_1, L_2.* These lines are the *focal lines* of the final

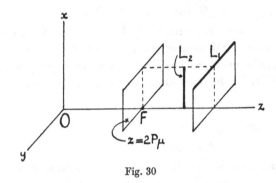

Fig. 30

bundle, but the general theory of focal lines is somewhat complicated by the approximations here employed. Fig. 30 shows these focal lines and their positions relative to the focal plane $z = 2P\mu$ for the case $Q' > 0$, $Q_{,,} > 0$, $P_{,} < 0$, $\sigma' > 0$. The diagram is not drawn to scale: actually the distance of L_1 from the axis of the instrument is of the order of σ', whereas the distances of L_1 and L_2 from the focal plane and from one another are much smaller, namely, of the order of σ'^2.

If $Q_{,,} = 0$, the two focal lines cut, and we get a point image, as indeed we know from earlier considerations.

A ray, as given by (15·24), cuts the focal plane $z = 2P\mu$ in the point satisfying

(15·26) $\begin{cases} x + P_{,}\sigma' + Q_{,}'\sigma'^3 = -2\sigma\sigma'^2(Q' + Q_{,,}), \\ y = -2\tau\sigma'^2 Q'. \end{cases}$

Thus all final rays inclined at a small angle θ to the axis of the instrument cut the focal plane in an ellipse,

(15·27) $\left(\dfrac{x + P_{,}\sigma' + Q_{,}'\sigma'^3}{Q' + Q_{,,}}\right)^2 + \left(\dfrac{y}{Q'}\right)^2 = 4\mu^2\theta^2\sigma'^4;$

its centre is at the point

(15·28) $$x = -P_{,}\sigma' - Q_{,}'\sigma'^3, \quad y = 0,$$

and its area is

(15·29) $$4\pi\mu^2\theta^2\sigma'^4 Q'(Q' + Q_{,,}).$$

The ellipse reduces to a straight line if

(15·30) $$Q' + Q_{,,} = 0 \quad \text{or} \quad Q' = 0,$$

in either of which cases one of the focal lines is in the focal plane.

The aberrational phenomenon present in the above system is known as *astigmatism*. The literal meaning of *astigmatism* refers to the failure of the instrument to form a point image.

Coma.

Let us now suppose that

(15·31) $$P = 4Q\mu^2, \quad Q_{,} \neq 0, \quad Q_{,,} = 0.$$

The equations (15·5) for the final rays may be written

(15·32) $$\begin{cases} x + P_{,}\sigma' + Q_{,}'\epsilon'\sigma' - \sigma(1 + \tfrac{1}{2}\epsilon\mu^{-2})(z\mu^{-1} - 2P - 2Q'\epsilon') \\ \qquad = -Q_{,}(\sigma'\epsilon + 2\epsilon_{,}\sigma), \\ y + P_{,}\tau' + Q_{,}'\epsilon'\tau' - \tau(1 + \tfrac{1}{2}\epsilon\mu^{-2})(z\mu^{-1} - 2P - 2Q'\epsilon') \\ \qquad = -Q_{,}(\tau'\epsilon + 2\epsilon_{,}\tau). \end{cases}$$

Let us, as before, take the xz plane parallel to the incident rays, so that (15·23) hold. Then (15·32) become

(15·33) $$\begin{cases} x + P_{,}\sigma' + Q_{,}'\sigma'^3 - \sigma(1 + \tfrac{1}{2}\epsilon\mu^{-2})(z\mu^{-1} - 2P - 2Q'\sigma'^2) \\ \qquad = -Q_{,}\sigma'(3\sigma^2 + \tau^2), \\ y - \tau(1 + \tfrac{1}{2}\epsilon\mu^{-2})(z\mu^{-1} - 2P - 2Q'\sigma'^2) = -2Q_{,}\sigma'\sigma\tau. \end{cases}$$

Let us examine the intersections of these rays with the plane

(15·34) $$z = 2\mu(P + Q'\sigma'^2):$$

by (15·20) this is the plane on which a point image would be formed if $Q_{,} = 0$. Shifting the origin to the point $\bar{O}$ with coordinates

(15·35) $$x = -P_{,}\sigma' - Q_{,}'\sigma'^3, \quad y = 0, \quad z = 0,$$

and denoting the new x by $\bar{x}$, we see that the ray with components σ, τ cuts the plane (15·34) at the point

(15·36) $$\bar{x} = -Q_{,}\sigma'(3\sigma^2 + \tau^2), \quad y = -2Q_{,}\sigma'\sigma\tau.$$

Let us introduce polar angles θ, ϕ, so that

(15·37) $\sigma = \mu \sin\theta \cos\phi, \quad \tau = \mu \sin\theta \sin\phi.$

Then the intersection is, to the third order,

(15·38)
$$\begin{cases} \bar{x} = -Q_i\sigma'\mu^2\theta^2(3\cos^2\phi + \sin^2\phi) \\ \qquad = -Q_i\sigma'\mu^2\theta^2(2+\cos 2\phi), \\ y = -2Q_i\sigma'\mu^2\theta^2 \sin\phi\cos\phi = -Q_i\sigma'\mu^2\theta^2 \sin 2\phi. \end{cases}$$

Those rays for which θ is constant cut the plane (15·34) in the circle

(15·39) $(\bar{x} + 2Q_i\sigma'\mu^2\theta^2)^2 + y^2 = (Q_i\sigma'\mu^2\theta^2)^2.$

The centre of this circle is at

$$\bar{x} = -2Q_i\sigma'\mu^2\theta^2, \quad y = 0,$$

and its radius is $|Q_i\sigma'\mu^2\theta^2|$. Fig. 31 shows the projection of this circle on the plane $z = 0$. It is evident that the tangents drawn from the origin $\bar{O}$ make angles of 30° with the x-axis.

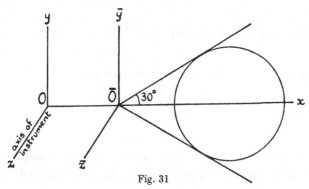

Fig. 31

(The figure is drawn for $P_i\sigma' < 0$, $Q_i\sigma' < 0$.) Letting θ vary, we see that *the illuminated portion of the plane* (15·34) *is a wedge or flare of angle* 60°, *having its vertex at the point whose coordinates are*

(15·40) $x = -P_i\sigma' - Q_i'\sigma'^3, \quad y = 0, \quad z = 2\mu(P + Q'\sigma'^2).$

This phenomenon is called *circular coma*. (Cf. § 11 for general, or elliptical, coma.) Since in (15·35) $P_i\sigma'$ is much greater than $Q_i'\sigma'^3$, we see that the vertex of the wedge points towards the axis of the instrument (as in Fig. 31) or away from it, according as P_iQ_i is positive or negative.

We have seen that a mirror, to be free from spherical aberration, must be paraboloidal, so that in (13·8) $S = \infty$. We have then from (13·11)

(15·41) $\qquad P_{,} = -\tfrac{1}{2}R, \quad Q_{,,} = 0, \quad Q_{,} = -\tfrac{1}{8}R.$

Thus *the paraboloidal mirror suffers from coma*, the vertex of the wedge pointing towards the axis, as in Fig. 31. Putting $\sigma' = \theta'$ in (15·39), we see that if the incident rays are inclined at an angle θ' to the axis, the final rays inclined at an angle θ to the axis cut the plane in a circle of radius $\tfrac{1}{8}R\theta'\theta^2$, the centre of the circle being at a distance $\tfrac{1}{4}R\theta'\theta^2$ from the vertex of the wedge. The vertex of the wedge is at a distance approximately $\tfrac{1}{2}R\theta'$ from the axis of the instrument. (R is the radius of curvature of the paraboloidal mirror at its vertex.)

Curvature of the image.

Let us suppose that an instrument is corrected for spherical aberration, astigmatism and coma, so that a point image is formed by any set of parallel incident rays. We have then

(15·42) $\qquad P = 4Q\mu^2, \quad Q_{,} = 0, \quad Q_{,,} = 0,$

and by (15·20) the point image is formed at

(15·43) $\qquad \begin{cases} x = -P_{,}\sigma' - Q_{,}'\epsilon'\sigma', \\ y = -P_{,}\tau' - Q_{,}'\epsilon'\tau', \\ z = 2\mu P + 2\mu Q'\epsilon'. \end{cases}$

Suppose now that there is an extended distant object, such as a planet, which we regard as lying at infinity. From each point of it there comes a family of parallel rays (σ', τ'), forming an image-point as given by (15·43). Varying σ', τ', we get an *image-surface*: this surface will in general be curved, so that a curved photographic plate would be necessary to obtain a sharp representation of the whole object.

The equation of the image-surface is approximately

(15·44) $\qquad z = 2\mu P + 2\mu Q'(x^2 + y^2)/P_{,}^2:$

thus *the radius of curvature ρ of the image-surface* (*counted positive when that surface has its concavity on the side $z = +\infty$*) is

(15·45) $\qquad \rho = P_{,}^2/(4\mu Q').$

The condition for a flat image (no curvature) is therefore

(15·46) $$Q' = 0.$$

In the case of an instrument with coma (conditions 15·31), there is not a point image, but the coordinates (15·40) for the vertex of the coma-flare are the same as (15·43) with $\tau' = 0$. If we regard the vertex of the flare as an image, we may speak of an image-surface and its curvature. For a paraboloidal mirror, as in (13·11),

(15·47) $$\rho = \tfrac{1}{2}R;$$

the radius of curvature of the image is half that of the mirror.

Distortion.

Let us now suppose that the instrument is corrected for spherical aberration, astigmatism, coma and curvature, so that

(15·48) $\quad P = 4Q\mu^2, \quad Q_{,} = 0, \quad Q_{,,} = 0, \quad Q' = 0.$

Then, by (15·43), the image point corresponding to σ', τ' is at

(15·49)
$$\begin{cases} x = -P_{,}\sigma' - Q'_{,}\epsilon'\sigma', \\ y = -P_{,}\tau' - Q'_{,}\epsilon'\tau', \\ z = 2\mu P_{.}. \end{cases}$$

Although a plane image of an extended object is formed, it may not be perfect; *distortion* may be present.

Let us first suppose that the object is plane, the plane being perpendicular to the axis of the instrument. Let the instrument be removed, and replaced by a screen perpendicular to the z-axis, the screen having an infinitesimal hole on the axis. This arrangement constitutes a "pin-hole camera", and an image of the object plane will be formed on any plane perpendicular to the axis behind the hole. Any pattern drawn on the object plane will be reproduced to scale on the image plane, all lengths being enlarged or reduced in the same ratio. This is an image without distortion.

If the object is not plane, we shall *define an image without distortion* as one formed in this way by projection through a point on the axis. The image formed by an optical instrument will be, by definition, free from distortion when it is a reproduction to scale of the image that would be formed by projection through a point.

Suppose then that there is an object point at infinity, the components of the rays from it being σ', τ'. Taking projection through the origin on to a plane $z = 1$, the ray σ', τ' gives the point

$$(15\cdot50) \qquad x = \sigma'/v', \quad y = \tau'/v'.$$

Denoting these coordinates by $\bar{x}$, $\bar{y}$, and using

$$(15\cdot51) \qquad \sigma'^2 + \tau'^2 + v'^2 = \mu'^2,$$

we see that to the third order of small quantities

$$(15\cdot52) \qquad \begin{cases} \sigma' = \eta'\mu'\bar{x}[1 - \tfrac{1}{2}(\bar{x}^2 + \bar{y}^2)], \\ \tau' = \eta'\mu'\bar{y}[1 - \tfrac{1}{2}(\bar{x}^2 + \bar{y}^2)], \end{cases}$$

where $\eta' = \pm 1$ according as the incident rays are in the positive or negative sense. By (15·49) and (15·52) the corresponding image-point *formed by the instrument* is at

$$(15\cdot53) \qquad \begin{cases} x = -\eta'\mu'P_{,}\bar{x}[1 + \tfrac{1}{2}(\bar{x}^2 + \bar{y}^2)(2Q'_{,}\mu'^2/P_{,} - 1)], \\ y = -\eta'\mu'P_{,}\bar{y}[1 + \tfrac{1}{2}(\bar{x}^2 + \bar{y}^2)(2Q'_{,}\mu'^2/P_{,} - 1)]. \end{cases}$$

The condition for a reproduction to scale, i.e. *the condition for no distortion*, is obviously

$$(15\cdot54) \qquad 2Q'_{,}\mu'^2 = P_{,}.$$

If this condition is not satisfied, the straight line $\bar{y} = \mathrm{const.}$ in the image by projection does not correspond to a straight line in the image formed by the instrument, but to a parabolic arc which curves away from $y = 0$ if

$$(15\cdot55) \qquad 2Q'_{,}\mu'^2/P_{,} - 1 > 0,$$

and toward $y = 0$ if

$$(15\cdot56) \qquad 2Q'_{,}\mu'^2/P_{,} - 1 < 0.$$

Also the straight line $\bar{x} = \mathrm{const.}$ corresponds to a parabolic arc, curving away from or towards $x = 0$ according as (15·55) or (15·56) holds.

The distortion corresponding to (15·55) is called *cushion* distortion and that corresponding to (15·56) is called *barrel* distortion, the names being suggested by the patterns corresponding to a rectangular grid of lines in the $\bar{x}\bar{y}$ plane (Fig. 32).

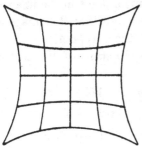

Fig. 32a. Cushion distortion

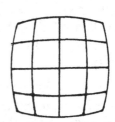

Fig. 32b. Barrel distortion

Collecting our results, we see that the conditions for the absence of spherical aberration, astigmatism, coma, curvature and distortion are

(15·57) $P = 4Q\mu^2$, $Q_{,} = Q_{,,} = Q' = 0$, $P_{,} = 2Q'_{,}\mu'^2$,

so that the T-function for such an instrument, perfect (to the order considered) for parallel incident rays, is

(15·58) $T = T^{(0)} + P'\epsilon' + Q''\epsilon'^2 + Q'_{,}\epsilon_{,}(2\mu'^2 + \epsilon') + Q\epsilon(4\mu^2 + \epsilon)$,

the four remaining constants being arbitrary.

16. The sine condition of Abbe.

The preceding theory is approximate, dealing with rays adjacent to the axis of the instrument. We proceed to establish a condition which must be satisfied no matter what the inclination of the rays to the axis may be, if the instrument fulfils certain conditions of imagery.

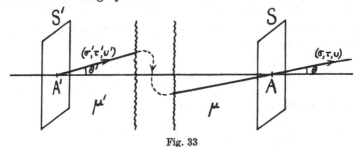

Fig. 33

Let S' be a plane object perpendicular to the axis of the instrument at A'. Let us suppose that all rays from each point of S' pass accurately through a single point in the image-space, so that an image-surface is formed, and that this image-surface of S' is a plane S perpendicular to the axis at A.

Let $W(x', y', z', \sigma, \tau)$ be the W-function of the instrument. Then by (6·10), (6·11) we have

(16·1) $$\sigma' = -W_{x'}, \quad \tau' = -W_{y'},$$

(16·2) $$x - z\sigma/v = -W_\sigma, \quad y - z\tau/v = -W_\tau.$$

Given x', y', z', σ, τ, the equations (16·1) determine σ', τ': (16·2) are the equations of the final ray with components σ, τ, v. Let x', y', z' be any point on S' and x, y, z the corresponding image-point on S. The origin is arbitrary: let us choose it at A, so that $z = 0$ and (16·2) read

(16·3) $$x = -W_\sigma, \quad y = -W_\tau.$$

Let us regard x', y', σ, τ as independent variables and x, y, σ', τ' as functions of them. Then by our assumption as to the nature of the image, x, y *are functions of x', y' only, independent of σ, τ.*

By partial differentiation of (16·1), (16·3) we have

(16·4)
$$\begin{cases} -W_{x'\sigma} = \dfrac{\partial \sigma'}{\partial \sigma}, & -W_{x'\tau} = \dfrac{\partial \sigma'}{\partial \tau}, & -W_{y'\sigma} = \dfrac{\partial \tau'}{\partial \sigma}, & -W_{y'\tau} = \dfrac{\partial \tau'}{\partial \tau}, \\[2mm] -W_{\sigma x'} = \dfrac{\partial x}{\partial x'}, & -W_{\tau x'} = \dfrac{\partial y}{\partial x'}, & -W_{\sigma y'} = \dfrac{\partial x}{\partial y'}, & -W_{\tau y'} = \dfrac{\partial y}{\partial y'}. \end{cases}$$

Hence

(16·5) $$\frac{\partial \sigma'}{\partial \sigma} = \frac{\partial x}{\partial x'}, \quad \frac{\partial \sigma'}{\partial \tau} = \frac{\partial y}{\partial x'}, \quad \frac{\partial \tau'}{\partial \sigma} = \frac{\partial x}{\partial y'}, \quad \frac{\partial \tau'}{\partial \tau} = \frac{\partial y}{\partial y'}.$$

Let the subscript 0 mean $x' = y' = 0$. Then from the symmetry of the instrument

(16·6) $$\left(\frac{\partial y}{\partial x'}\right)_0 = \left(\frac{\partial x}{\partial y'}\right)_0 = 0, \quad \left(\frac{\partial x}{\partial x'}\right)_0 = \left(\frac{\partial y}{\partial y'}\right)_0 = m,$$

m being the magnification. Hence

(16·7) $$\left(\frac{\partial \sigma'}{\partial \tau}\right)_0 = \left(\frac{\partial \tau'}{\partial \sigma}\right)_0 = 0, \quad \left(\frac{\partial \sigma'}{\partial \sigma}\right)_0 = \left(\frac{\partial \tau'}{\partial \tau}\right)_0 = m.$$

Therefore

(16·8) $(\sigma')_0 = m\sigma + a, \quad (\tau')_0 = m\tau + b,$

where a, b are constants. But $(\sigma')_0 = 0$ if $\sigma = 0$ and $(\tau')_0 = 0$ if $\tau = 0$. Hence $a = b = 0$, and so for corresponding rays through A', A we have

(16·9) $\sigma' = m\sigma, \quad \tau' = m\tau.$

Thus *if an exact plane image S of the plane S' is formed, the inclinations θ', θ to the axis of the instrument of corresponding rays through A' and A must satisfy*

(16·10) $\mu' \sin \theta' = m\mu \sin \theta,$

μ', μ *being the refractive indices of the initial and final media.* This is known as *the sine condition of Abbe.*

This result may also be established by means of the point-characteristic V, perhaps more directly. From the assumed property of exact imagery and Fermat's principle, all the rays joining an assigned point on the plane through A' to its image have the same optical length V, which is a function of x', y' only. By symmetry V must be a maximum or minimum when $x' = y' = 0$, and hence $\delta V = 0$ for any infinitesimal displacement off the axis at A'. Therefore, by (5·8),

(16·11) $\delta V = \sigma\, \delta x + \tau\, \delta y - \sigma'\, \delta x' - \tau'\, \delta y' = 0,$

where σ', τ' are the components of any ray through A' and σ, τ the components of the corresponding final ray through A; $\delta x'$, $\delta y'$ is an *arbitrary* displacement in the plane at A' and δx, δy the *corresponding* displacement at A. Combining (16·11) with (16·6), we deduce (16·9), and hence the sine condition (16·10) follows. This method is immediately applicable to the more general case in which S', S are not planes, but surfaces of revolution about the axis of the instrument: in that case also (16·10) holds.

17. Calculation of T for a thin system.

Consider an instrument of revolution (Fig. 34) in which surfaces $S_i (i = 1, 2, ..., n)$ separate media of refractive indices $\mu_0, \mu_1, ..., \mu_n$. Let the equation of S_i be approximately

(17·1) $z = v_i + \tfrac{1}{2} r_i(x^2 + y^2) + \tfrac{1}{4} s_i(x^2 + y^2)^2,$

where v_i, r_i, s_i are constants, r_i being the reciprocal of the radius of curvature, counted positive when the surface is concave towards $z = +\infty$. The notation has been slightly altered from that of (12·1): we note that the condition for a spherical surface is now

(17·2) $$s_i = \tfrac{1}{2}r_i^3.$$

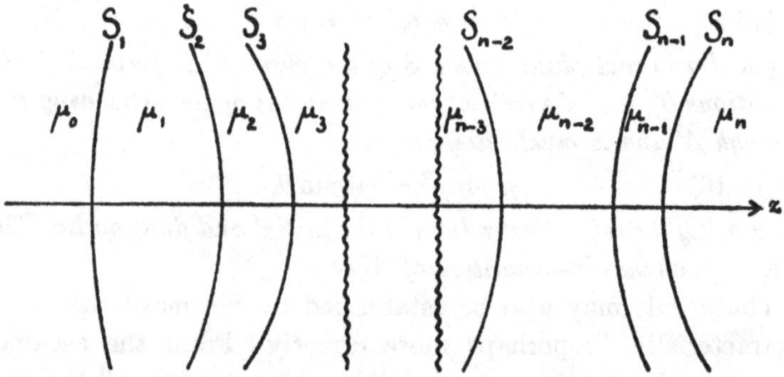

Fig. 34

A considerable mathematical simplification results from assuming $v_i = 0 \ (i = 1, 2, \ldots, n)$: although it is physically impossible to bring the vertices of the surfaces into coincidence without breaking the refracting material, we shall assume that this condition is satisfied. The system so obtained is called a *thin system* in a technical sense. The behaviour of an actual instrument will approach more and more closely to that of a thin system the smaller the distances between the vertices.

We shall denote the components in the several media by

$$\sigma_0, \tau_0; \ \sigma_1, \tau_1; \ \ldots; \ \sigma_n, \tau_n.$$

For any quantity ψ we shall denote the backward difference by

(17·3) $$\nabla_i \psi = \psi_i - \psi_{i-1}.$$

Thus, for example, $\nabla_i \mu$ is the increment in refractive index on crossing the surface S_i.

With the necessary changes in notation, (12·22) gives for the T-function for the media separated by S_i, to the fourth order,

(17·4) $$T_{i-1,i} = \tfrac{1}{2}r_i^{-1}(\nabla_i\mu)^{-1}[(\nabla_i\sigma)^2 + (\nabla_i\tau)^2]$$
$$\times \{1 + \tfrac{1}{2}(\nabla_i\mu)^{-1}\nabla_i[\mu^{-1}(\sigma^2 + \tau^2)]\}$$
$$- \tfrac{1}{8}s_i r_i^{-3}(\nabla_i\mu)^{-2}[(\nabla_i\sigma)^2 + (\nabla_i\tau)^2]\}.$$

This may be written

$$(17\cdot5) \quad \begin{cases} T_{i-1,i} = T^{(2)}_{i-1,i} + T'^{(4)}_{i-1,i} + T''^{(4)}_{i-1,i}, \\ T^{(2)}_{i-1,i} = \tfrac{1}{2}r_i^{-1}(\nabla_i\mu)^{-1}\left[(\nabla_i\sigma)^2 + (\nabla_i\tau)^2\right], \\ T'^{(4)}_{i-1,i} = \tfrac{1}{4}r_i^{-1}(\nabla_i\mu)^{-2}\left[(\nabla_i\sigma)^2 + (\nabla_i\tau)^2\right]\nabla_i[\mu^{-1}(\sigma^2+\tau^2)], \\ T''^{(4)}_{i-1,i} = -\tfrac{1}{8}s_i r_i^{-4}(\nabla_i\mu)^{-3}\left[(\nabla_i\sigma)^2 + (\nabla_i\tau)^2\right]^2. \end{cases}$$

The T-function for the complete instrument is

$$(17\cdot6) \quad \begin{cases} T = T^{(2)} + T'^{(4)} + T''^{(4)}, \\ T^{(2)} = \sum_{i=1}^{n} T^{(2)}_{i-1,i}, \\ T'^{(4)} = \sum_{i=1}^{n} T'^{(4)}_{i-1,i}, \\ T''^{(4)} = \sum_{i=1}^{n} T''^{(4)}_{i-1,i}. \end{cases}$$

The intermediate components σ_i, τ_i $(i = 1, 2, ..., n-1)$ are to be eliminated, in accordance with the approximate method justified in § 13, by means of

$$(17\cdot7) \quad \frac{\partial T^{(2)}}{\partial\sigma_i} = 0, \quad \frac{\partial T^{(2)}}{\partial\tau_i} = 0, \qquad (i = 1, 2, ..., n-1).$$

Now

$(17\cdot8)$

$$\begin{cases} T^{(2)} = \tfrac{1}{2}\sum_{j=1}^{n} (r_j\nabla_j\mu)^{-1}\{(\sigma_j - \sigma_{j-1})^2 + (\tau_j - \tau_{j-1})^2\}, \\ \dfrac{\partial T^{(2)}}{\partial\sigma_i} = (r_i\nabla_i\mu)^{-1}(\sigma_i - \sigma_{i-1}) - (r_{i+1}\nabla_{i+1}\mu)^{-1}(\sigma_{i+1} - \sigma_i) = 0. \end{cases}$$

Hence

$$(17\cdot9) \quad \begin{cases} \dfrac{\nabla_i\sigma}{r_i\nabla_i\mu} = \dfrac{\nabla_{i+1}\sigma}{r_{i+1}\nabla_{i+1}\mu} = C, \\ \dfrac{\nabla_i\tau}{r_i\nabla_i\mu} = \dfrac{\nabla_{i+1}\tau}{r_{i+1}\nabla_{i+1}\mu} = D, \end{cases}$$

where C, D are independent of i. These fractions are *invariants*.

Let us put

$$(17\cdot10) \quad F_i^{-1} = \sum_{j=1}^{i} r_j\nabla_j\mu, \quad (i = 1, 2, ..., n).$$

Then by (17·9)

(17·11)

$$\left\{ \begin{aligned} \sigma_i - \sigma_0 &= \sum_{j=1}^{i} \nabla_j \sigma = C \sum_{j=1}^{i} r_j \nabla_j \mu = C F_i^{-1} \\ \tau_i - \tau_0 &= \sum_{j=1}^{i} \nabla_j \tau = D \sum_{j=1}^{i} r_j \nabla_j \mu = D F_i^{-1} \end{aligned} \right\} \quad (i = 1, \dots, n),$$

and so

(17·12) $\sigma_n - \sigma_0 = C F^{-1}, \quad \tau_n - \tau_0 = D F^{-1},$

where we define

(17·13) $$F^{-1} = F_n^{-1} = \sum_{j=1}^{n} r_j \nabla_j \mu.$$

By (17·11), (17·12) we have

(17·14) $\left\{ \begin{aligned} \sigma_i - \sigma_0 &= (\sigma_n - \sigma_0) F F_i^{-1} \\ \tau_i - \tau_0 &= (\tau_n - \tau_0) F F_i^{-1} \end{aligned} \right\} \quad (i = 1, \dots, n-1).$

These equations give the intermediate components in terms of the initial and final components.

By (17·9), (17·12) we have

(17·15) $(\nabla_i \sigma)^2 + (\nabla_i \tau)^2 = (r_i \nabla_i \mu)^2 F^2 \{ (\sigma_n - \sigma_0)^2 + (\tau_n - \tau_0)^2 \}.$

Hence by (17·8), (17·13) *we have for the T-function for the complete instrument to the second order*

(17·16) $T^{(2)} = \frac{1}{2} F \{ (\sigma_n - \sigma_0)^2 + (\tau_n - \tau_0)^2 \}.$

By (17·5), (17·9), (17·12)

(17·17) $\left\{ \begin{aligned} T'^{(4)}_{i-1,i} &= \tfrac{1}{4} r_i F^2 \{ (\sigma_n - \sigma_0)^2 + (\tau_n - \tau_0)^2 \} \nabla_i \phi, \\ \phi_i &= \mu_i^{-1} (\sigma_i^2 + \tau_i^2). \end{aligned} \right.$

Now

(17·18)

$$\left\{ \begin{aligned} \sum_{i=1}^{n} r_i \nabla_i \phi &= r_1(\phi_1 - \phi_0) + r_2(\phi_2 - \phi_1) + \dots + r_n(\phi_n - \phi_{n-1}), \\ &= -r_1 \phi_0 + r_n \phi_n - (r_2 - r_1) \phi_1 - \dots - (r_n - r_{n-1}) \phi_{n-1}, \\ &= -r_1 \phi_0 + r_n \phi_n - \sum_{i=1}^{n-1} \phi_i \nabla_{i+1} r. \end{aligned} \right.$$

Hence

(17·19). $$T'^{(4)} = \sum_{i=1}^{n} T''^{(4)}_{i-1,i} = \tfrac{1}{4}F^2\{(\sigma_n-\sigma_0)^2+(\tau_n-\tau_0)^2\}$$
$$\times\{r_n\phi_n - r_1\phi_0 - \sum_{i=1}^{n-1}\phi_i\nabla_{i+1}r\},$$

in which the ϕ's are to be replaced by the following expressions, obtained on substitution from (17·14) in the definitions (17·17):

(17·20)

$$\begin{cases} \phi_i = \mu_i^{-1}\{[\sigma_0+(\sigma_n-\sigma_0)\,FF_i^{-1}]^2+[\tau_0+(\tau_n-\tau_0)\,FF_i^{-1}]^2\}, \\ \qquad\qquad (i=1,2,...,n-1), \\ \phi_0 = \mu_0^{-1}(\sigma_0^2+\tau_0^2), \quad \phi_n = \mu_n^{-1}(\sigma_n^2+\tau_n^2). \end{cases}$$

Thus (17·19) *expresses* $T'^{(4)}$ *as a function of the initial and final components.*

Lastly, by (17·5), (17·9), (17·12),

(17·21) $$T''^{(4)} = -\tfrac{1}{4}F^4\{(\sigma_n-\sigma_0)^2+(\tau_n-\tau_0)^2\}^2\sum_{i=1}^{n} s_i\nabla_i\mu.$$

Collecting our results and writing σ', τ' for the initial components and σ, τ for the final components, *we have in the notation of* (13·3) *for the T-function of any thin system of revolution at the origin, to the fourth order inclusive,*

(17·22)

$$\begin{cases} T = T^{(2)} + T^{(4)}, \\ T^{(2)} = \tfrac{1}{2}F(\epsilon'-2\epsilon,+\epsilon), \\ T^{(4)} = \tfrac{1}{4}F^2(\epsilon'-2\epsilon,+\epsilon)\,[r_n\mu_n^{-1}\epsilon - r_1\mu_0^{-1}\epsilon' - \sum_{i=1}^{n-1}\mu_i^{-1}\nabla_{i+1}r \\ \qquad \times\{\epsilon'(1-FF_i^{-1})^2+2\epsilon,FF_i^{-1}(1-FF_i^{-1})+\epsilon F^2F_i^{-2}\}] \\ \qquad\qquad\qquad - \tfrac{1}{4}F^4(\epsilon'-2\epsilon,+\epsilon)^2\sum_{i=1}^{n}s_i\nabla_i\mu. \end{cases}$$

If we write as usual

(17·23) $$T = P'\epsilon' + P,\epsilon, + P\epsilon + Q''\epsilon'^2$$
$$ + Q_{,,}\epsilon_{,}^2 + Q\epsilon^2 + Q_{,}'\epsilon'\epsilon, + Q'\epsilon'\epsilon + Q,\epsilon,\epsilon,$$

we have, by comparison of coefficients,

(17·24)

$$
\begin{cases}
P' = P = \tfrac{1}{2}F, \quad P_, = -F, \\[2mm]
Q = \tfrac{1}{4}F^2(r_n\mu_n^{-1} - F^2\sum_{i=1}^{n-1}F_i^{-2}\mu_i^{-1}\nabla_{i+1}r) - \tfrac{1}{4}F^4\sum_{i=1}^{n}s_i\nabla_i\mu, \\[2mm]
Q_, = -\tfrac{1}{2}F^2[r_n\mu_n^{-1} + \sum_{i=1}^{n-1}FF_i^{-1}(1 - 2FF_i^{-1})\mu_i^{-1}\nabla_{i+1}r] \\[2mm]
\qquad\qquad\qquad\qquad\qquad\qquad + F^4\sum_{i=1}^{n}s_i\nabla_i\mu, \\[2mm]
\qquad\qquad (F_i^{-1} = \sum_{j=1}^{i}r_j\nabla_j\mu,\ F = F_n).
\end{cases}
$$

If $\mu_0 = \mu_n = 1$, we have by (14·11) for the focal points

(17·25) $z' = -2P' = -F, \quad z = 2P = F,$

while the nodal and principal points lie at $z = 0$; the focal length of the instrument is by (14·28)

(17·26) $$F = (\sum_{i=1}^{n}r_i\nabla_i\mu)^{-1}.$$

F^{-1} is called the *power*. If we define the power of the pair of media separated by S_i to be $r_i\nabla_i\mu$, then the power of the whole instrument is the sum of the powers of the consecutive pairs of media.

We note that T contains the factor $\epsilon' - 2\epsilon_, + \epsilon$: hence we may write

(17·27) $T^{(4)} = (\epsilon' - 2\epsilon_, + \epsilon)(A\epsilon' + B\epsilon_, + C\epsilon),$

and we shall have

(17·28) $$\begin{cases} Q'' = A, \quad Q_{,,} = -2B, \quad Q = C, \\ Q'_, = B - 2A, \quad Q' = C + A, \quad Q_, = -2C + B. \end{cases}$$

Of the Q's only three are independent: we have in fact

(17·29) $Q'_, = -\tfrac{1}{2}Q_{,,} - 2Q'', \quad Q' = Q + Q'', \quad Q_, = -2Q - \tfrac{1}{2}Q_{,,}.$

We recall from (15·11), (15·16) that the conditions for the formation of point images are

(17·30) $P = 4Q\mu_n^2, \quad Q_, = 0, \quad Q_{,,} = 0.$

Since $P = \frac{1}{2}F \neq 0$, if the focal length is not to vanish, the first of (17·30) demands $Q \neq 0$, and hence (17·30) are incompatible with the last of (17·29). Thus *it is impossible to correct a thin instrument simultaneously for spherical aberration, astigmatism and coma.*

18. Aberrations of a thin lens.

Let us now consider a thin lens *in vacuo*, bounded by the spherical surfaces

$$(18\cdot1) \quad \begin{cases} S_1: & z = \frac{1}{2}r_1(x^2+y^2) + \frac{1}{4}s_1(x^2+y^2)^2, \quad s_1 = \frac{1}{2}r_1^3, \\ S_2: & z = \frac{1}{2}r_2(x^2+y^2) + \frac{1}{4}s_2(x^2+y^2)^2, \quad s_2 = \frac{1}{2}r_2^3; \end{cases}$$

r_1, r_2 are the reciprocals of the radii. The figure shows the case of

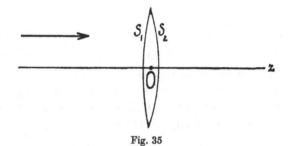

Fig. 35

a convex lens, $r_1 > 0$, $r_2 < 0$, but the argument to be developed applies to all signs.

We are to put in the results of § 17

$$(18\cdot2) \quad \mu_0 = \mu_2 = 1, \quad \mu_1 = \mu, \quad \nabla_1\mu = \mu - 1, \quad \nabla_2\mu = 1 - \mu,$$

μ being the refractive index of the lens. For the focal length F we have

$$(18\cdot3) \qquad F^{-1} = r_1\nabla_1\mu + r_2\nabla_2\mu = (\mu - 1)(r_1 - r_2),$$

and so by (17·24) we have

$$(18\cdot4) \quad P' = P = -\frac{1}{2}P, = \frac{1}{2}F = \frac{1}{2}(\mu-1)^{-1}(r_1 - r_2)^{-1}.$$

Also

$$(18\cdot5) \qquad\qquad F_1^{-1} = r_1\nabla_1\mu = r_1(\mu - 1).$$

Application of (17·24) with $n = 2$ gives

$$(18\cdot6) \begin{cases} Q = \tfrac{1}{4}F^4[r_2 F^{-2} - \mu^{-1}(r_2 - r_1) F_1^{-2} - (s_1 - s_2)(\mu - 1)] \\ \quad = \tfrac{1}{4}F^2[r_2 + \mu^{-1} r_1^2 (r_1 - r_2)^{-1} \\ \qquad\qquad - \tfrac{1}{2}(\mu - 1)^{-1}(r_1^2 + r_1 r_2 + r_2^2)(r_1 - r_2)^{-1}], \\ Q_, = -\tfrac{1}{2}F^2[r_2 + F F_1^{-1}(1 - 2 F F_1^{-1})\mu^{-1}(r_2 - r_1)] \\ \qquad\qquad + F^4(s_1 - s_2)(\mu - 1) \\ \quad = -\tfrac{1}{2}F^2[r_2 + \mu^{-1} r_1(r_1 + r_2)(r_1 - r_2)^{-1} \\ \qquad\qquad - (\mu - 1)^{-1}(r_1^2 + r_1 r_2 + r_2^2)(r_1 - r_2)^{-1}]. \end{cases}$$

Hence

$$(18\cdot7) \quad 4Q - P = F^2[r_2 - \tfrac{1}{2}(\mu - 1)(r_1 - r_2) + \mu^{-1} r_1^2 (r_1 - r_2)^{-1} \\ - \tfrac{1}{2}(\mu - 1)^{-1}(r_1^2 + r_1 r_2 + r_2^2)(r_1 - r_2)^{-1}].$$

Thus by (15·11) *the condition for the absence of spherical aberration is*

$$(18\cdot8) \quad (4Q - P)(r_1 - r_2) F^{-2} \equiv r_2(r_1 - r_2) - \tfrac{1}{2}(\mu - 1)(r_1 - r_2)^2 \\ + \mu^{-1} r_1^2 - \tfrac{1}{2}(\mu - 1)^{-1}(r_1^2 + r_1 r_2 + r_2^2) = 0.$$

Given μ, this is a quadratic equation for the ratio r_1/r_2. However, since $\mu > 1$ necessarily, the roots prove to be imaginary, so that it is impossible to avoid spherical aberration for a single thin lens.

We have also from (18·6) and (17·29)

$$(18\cdot9) \quad \begin{cases} 4Q + Q_, = \tfrac{1}{2}F^2[r_2 + \mu^{-1} r_1], \\ 4Q - Q_{,,} = 2(4Q + Q_,), \end{cases}$$

so that $Q_,$, $Q_{,,}$ are easily calculated when Q has been found from (18·6).

19. Chromatic aberrations.

In the preceding work we have treated the refractive index of a medium as a constant. Actually it depends on the colour or frequency of the light employed. Thus our investigations up to this point must (except in the case of reflections) be regarded as applying to light of a single colour (monochromatic light). The phenomena arising from variation of refractive index with colour are known as phenomena of *dispersion*.

Since the T-function of an instrument depends on the refractive indices of the media involved, it will depend on the colour of the light. Thus we should write

(19·1) $T = T(\sigma', \tau', \sigma, \tau, \chi),$

where χ is some number which specifies the colour of the light. We might take for χ the wave-length of the light *in vacuo*, or its frequency.

In an instrument involving reflections only, dispersion is entirely absent. This is obvious from the fact that the law of reflection does not depend on the refractive index, but it may also be seen by consideration of the T-function. If the reflections take place *in vacuo*, T will be independent of χ. If they take place in a medium of index μ, T will be of the form $\mu T'$, where T' is a function of direction cosines only, and the factor μ will disappear in the equations of the rays.

In the instrument of revolution, for which T has the approximate form (13·6), (13·7), the coefficients P', $P_{,}$, P, Q'', ... will depend on χ. Hence we are not to expect that the absence of spherical aberration, for example, for one colour will imply its absence for other colours. More serious than this, however, the dependence of T on χ makes itself felt even in the approximate theory based on the second-order terms in T, and leads to *chromatic aberrations* much more important than those arising from the dependence of the Q's on χ. The first-order imagery of § 14 will in general be different for different colours, because the quantities $\mu'P', \mu'P_{,}, \mu P_{,}, \mu P$

(and hence the cardinal points) will be different. We shall confine our attention to instruments *in vacuo*, so that $\mu' = \mu = 1$; the quantities determining the cardinal points are then P', $P_{,}$, P.

We cannot design a refracting instrument to make P', $P_{,}$, P independent of χ. But, with sufficient parameters (viz. refractive indices, curvatures and positions of surfaces) at our disposal, we can make these quantities take the same values for specified values of χ, i.e. we can eliminate chromatic aberration for specified colours. Actually it is usual in practice to limit the correction to two values of χ, say χ_1, χ_2 (corresponding to the C and F lines of the sodium spectrum). Using Δ_χ to indicate an increment on passing from χ_1 to χ_2, the conditions for achromatism for these two colours are

(19·2) $\Delta_\chi P' = 0, \quad \Delta_\chi P_, = 0, \quad \Delta_\chi P = 0.$

For a thin combination *in vacuo* these reduce by (17·24) to the single condition

(19·3) $\Delta_\chi F^{-1} = \Delta_\chi (\sum\limits_{i=1}^{n} r_i \nabla_i \mu) = 0,$

or

(19·4) $\sum\limits_{i=1}^{n} r_i \nabla_i (\Delta_\chi \mu) = 0.$

For a *single thin lens* with refractive index μ, this condition reads

(19·5) $(r_1 - r_2) \Delta_\chi \mu = 0,$

which cannot be satisfied except by the trivial solution $r_1 = r_2$. For a *thin double lens* with curvatures r_1, r_2, r_3, r_4 and refractive indices μ_1, μ_3, the condition for achromatism is

(19·6) $(r_1 - r_2) \Delta_\chi \mu_1 + (r_3 - r_4) \Delta_\chi \mu_3 = 0.$

The *dispersive power* of a medium is conventionally defined as

(19·7) $$D = \frac{\Delta_\chi \mu}{\bar{\mu} - 1},$$

where $\bar{\mu}$ is the refractive index for some colour fixed conventionally (the sodium D-line). Thus (19·6) reads

(19·8) $D_1 (r_1 - r_2) (\bar{\mu}_1 - 1) + D_3 (r_3 - r_4) (\bar{\mu}_3 - 1) = 0,$

or

(19·9) $D_1/F_1 + D_3/F_3 = 0,$

where D_1, D_3 are the dispersive powers of the lenses and F_1, F_3 their focal lengths for the colour corresponding to the index $\bar{\mu}$.

It is easily seen that for a general thin system of lenses, the condition for achromatism is

(19·10) $\Sigma(D/F) = 0,$

where the summation extends over all the lenses, D and F being respectively the dispersive power and the focal length of a lens.

CHAPTER V

HETEROGENEOUS ISOTROPIC MEDIA

20. Fermat's principle.

The media previously considered were *homogeneous* and *isotropic*, in the sense that the optical properties were the same at all points (homogeneity) and the same for all directions at each point (isotropy). The most general medium is *heterogeneous* and *anisotropic*, but we shall confine our attention to media which are *heterogeneous* and *isotropic*.

It is assumed that in a heterogeneous isotropic medium there is a *velocity of propagation* at each point, in general variable from point to point: we may write it

(20·1) $$v = v(x, y, z),$$

x, y, z being rectangular Cartesian coordinates. The function v will also depend on the colour of the light, but it will be unnecessary to indicate this dependence. The *refractive index* is defined as

(20·2) $$\mu = c/v,$$

c being the velocity of light *in vacuo*: μ varies from point to point, and so the heterogeneous isotropic medium may be called a *medium of variable refractive index*. It is assumed that in each medium μ is continuous and possesses continuous partial derivatives.

If we draw any curve C, joining points A' and A, a point moving from A' to A along C and having in each position the assigned velocity v, will pass from A' to A in time

(20·3) $$t = \int_{A'}^{A} ds/v = c^{-1} \int_{A'}^{A} \mu \, ds.$$

We define the *optical length* of C to be $\int_{A'}^{A} \mu \, ds$, as in § 2.

We shall accept as a basis for our theory *Fermat's principle* in the following form: *the actual ray along which light travels from A' to A has a stationary optical length when compared with adjacent* *curves joining A' and A*. This is, of course, equivalent to saying that the time taken by the light to travel from A' to A has a stationary value.

We shall now find the differential equations of the rays in a single medium. For any curve C joining A' and A with equations

$$(20\cdot4) \qquad x = x(u), \quad y = y(u), \quad z = z(u),$$

u being any parameter, the optical length is

$$(20\cdot5) \qquad L = \int_{u_1}^{u_2} \mu(x, y, z)\,(\dot{x}^2 + \dot{y}^2 + \dot{z}^2)^{\frac{1}{2}}\, du,$$

where $u = u_1$ at A' and $u = u_2$ at A and $\dot{x} = dx/du$, $\dot{y} = dy/du$, $\dot{z} = dz/du$. This may be written

$$(20\cdot6) \qquad \begin{cases} L = \displaystyle\int_{u_1}^{u_2} w\, du, \\[2mm] w = \mu(x, y, z)\,(\dot{x}^2 + \dot{y}^2 + \dot{z}^2)^{\frac{1}{2}}, \end{cases}$$

w being a function of $x, y, z, \dot{x}, \dot{y}, \dot{z}$. Let us take a set of adjacent curves, each described by equations of the form $(20\cdot4)$, the parameter u running between the same terminal values on all the curves. Then the variation of L on passing from one of these curves to its neighbour is

$$(20\cdot7) \qquad \begin{aligned} \delta L &= \int_{u_1}^{u_2} \delta w\, du \\ &= \int_{u_1}^{u_2} \left(\Sigma \frac{\partial w}{\partial \dot{x}} \delta \dot{x} + \Sigma \frac{\partial w}{\partial x} \delta x \right) du, \end{aligned}$$

Σ indicating a sum of terms obtained by changing $x \to y \to z$, and $\delta \dot{x}, \delta \dot{y}, \delta \dot{z}, \delta x, \delta y, \delta z$ being infinitesimal increments obtained in passing from a point on one curve to the point on its neighbour with the same value of u. Then

$$(20\cdot8) \qquad \delta \dot{x} = \frac{d}{du} \delta x; \quad \delta \dot{y} = \frac{d}{du} \delta y; \quad \delta \dot{z} = \frac{d}{du} \delta z,$$

and hence integration by parts gives

$$(20\cdot9) \quad \delta L = \left[\Sigma \frac{\partial w}{\partial \dot{x}} \delta x \right]_{u_1}^{u_2} - \int_{u_1}^{u_2} \Sigma \left(\frac{d}{du} \frac{\partial w}{\partial \dot{x}} - \frac{\partial w}{\partial x} \right) \delta x\, du.$$

The first part vanishes if the curves have common end-points, for then $\delta x = \delta y = \delta z = 0$ at those points. If the curve C is the natural ray from A' to A, the remaining integral must vanish for values of δx, δy, δz arbitrary along C save for the condition of vanishing at A' and A. Hence it follows that, *by virtue of Fermat's principle, a ray satisfies the differential equations*

(20·10)
$$\begin{cases} \dfrac{d}{du}\dfrac{\partial w}{\partial \dot{x}} - \dfrac{\partial w}{\partial x} = 0, \\[2mm] \dfrac{d}{du}\dfrac{\partial w}{\partial \dot{y}} - \dfrac{\partial w}{\partial y} = 0, \\[2mm] \dfrac{d}{du}\dfrac{\partial w}{\partial \dot{z}} - \dfrac{\partial w}{\partial z} = 0, \end{cases}$$

these being in fact Euler's equations for the extremals of $\int w\,du$.

Substituting for w from (20·6) we have

(20·11) $$\frac{d}{du}\frac{\partial w}{\partial \dot{x}} - \frac{\partial w}{\partial x} = \frac{d}{du}\left[\frac{\mu \dot{x}}{(\dot{x}^2 + \dot{y}^2 + \dot{z}^2)^{\frac{1}{2}}}\right] - \frac{\partial \mu}{\partial x}(\dot{x}^2 + \dot{y}^2 + \dot{z}^2)^{\frac{1}{2}},$$

and similar forms for the other two expressions in (20·10).

The parameter u along the ray C is still arbitrary. If we put $u = s$, the arc length of C, we have

(20·12) $$\dot{x}^2 + \dot{y}^2 + \dot{z}^2 = 1$$

along C. Hence, by (20·11), the equations (20·10) become

(20·13)
$$\begin{cases} \dfrac{d}{ds}\left(\mu \dfrac{dx}{ds}\right) - \dfrac{\partial \mu}{\partial x} = 0, \\[2mm] \dfrac{d}{ds}\left(\mu \dfrac{dy}{ds}\right) - \dfrac{\partial \mu}{\partial y} = 0, \\[2mm] \dfrac{d}{ds}\left(\mu \dfrac{dz}{ds}\right) - \dfrac{\partial \mu}{\partial z} = 0. \end{cases}$$

If we take for u a parameter defined by

(20·14) $$u = \int_{A'} ds/\mu,$$

we have $du = ds/\mu$, and so the equations of a ray may be written

$$(20\cdot15) \quad \begin{cases} \dfrac{d^2x}{du^2} = \dfrac{\partial}{\partial x}\,(\tfrac{1}{2}\mu^2), \\[2mm] \dfrac{d^2y}{du^2} = \dfrac{\partial}{\partial y}\,(\tfrac{1}{2}\mu^2), \\[2mm] \dfrac{d^2z}{du^2} = \dfrac{\partial}{\partial z}\,(\tfrac{1}{2}\mu^2). \end{cases}$$

Thus we have in (20·10), (20·13) *and* (20·15) *three different forms for the equations of the rays.*

Denoting by $\mathbf{i}$ the unit tangent vector to the ray, so that, by the first Frenet formula,

$$(20\cdot16) \qquad d\mathbf{i}/ds = \mathbf{j}/\rho,$$

where $\mathbf{j}$ is the unit principal normal, drawn to the concave side of the projection of the ray on its osculating plane, and ρ is the radius of curvature (always positive), the equations (20·13) read in vector form

$$(20\cdot17) \qquad \frac{d}{ds}\,(\mu\mathbf{i}) = \operatorname{grad}\mu,$$

or

$$(20\cdot18) \qquad \frac{d\mu}{ds}\,\mathbf{i} + \mu\mathbf{j}/\rho = \operatorname{grad}\mu.$$

Thus the gradient of the refractive index lies in the osculating plane of the ray. Also, operating on (20·18) with $\mathbf{j}\,.$, we get

$$(20\cdot19) \quad \mu/\rho = \mathbf{j}\,.\operatorname{grad}\mu = \partial\mu/\partial n, \quad \rho^{-1} = \partial(\log\mu)/\partial n,$$

where $\partial/\partial n$ indicates differentiation along the principal normal, ∂n being an element of length of this normal. Since ρ is positive by definition, $\partial\mu/\partial n$ is positive. Thus the refractive index increases as we go along the principal normal, or, in other words, *the ray bends toward the region of higher refractive index* (Fig. 36).

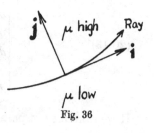
Fig. 36

Let us consider the case where the medium consists of *parallel planes of equal refractive index*. Taking the z-axis perpendicular to these planes, we have $\mu = \mu(z)$. Then integration of the first two of (20·13) gives

(20·20) $\mu\, dx/ds = a, \quad \mu\, dy/ds = b,$

where a and b are constants. Hence

(20·21) $(dz/ds)^2 = 1 - (dx/ds)^2 - (dy/ds)^2 = 1 - (a^2 + b^2)/\mu^2.$

We note that

(20·22) $dy/dx = b/a,$

and hence the *projection of each ray on $z = $ const. is a straight line*, as indeed we might expect from symmetry.

From (20·20), (20·21) we have

(20·23) $(dz/dx)^2 = (\mu^2 - a^2 - b^2)/a^2, \quad (dz/dy)^2 = (\mu^2 - a^2 - b^2)/b^2,$

which give x and y as functions of z by quadrature. Hence *we have for a ray in a stratified medium $\mu = \mu(z)$ the integrated equations*

(20·24) $x = a \int \dfrac{\pm\, dz}{(\mu^2 - a^2 - b^2)^{\frac{1}{2}}}, \quad y = b \int \dfrac{\pm\, dz}{(\mu^2 - a^2 - b^2)^{\frac{1}{2}}},$

the ambiguous signs corresponding to those occurring when the roots of (20·23) are taken. To avoid confusion arising from these ambiguities and to get a general idea of the behaviour of the rays, we may proceed as follows. Since the left-hand sides in (20·23) cannot be negative, the ray cannot leave the region for which

(20·25) $\mu \geqslant (a^2 + b^2)^{\frac{1}{2}}.$

The constants a and b are determined by the initial point and direction of the ray, and the right-hand side of (20·25) has a simple meaning. If θ denotes the inclination of the ray to the z-axis, we have in general

(20·26) $a^2 + b^2 = \mu^2[(dx/ds)^2 + (dy/ds)^2] = \mu^2 \sin^2\theta,$

and so (20·25) may be written

(20·27) $\mu - \mu' \sin\theta' \geqslant 0,$

the accents denoting initial values. The medium may be divided into layers in which the sign of the left-hand side of (20·27) is alternately positive and negative. The initial point lies in a layer for which this quantity is positive, and the ray cannot leave this layer, which is bounded by planes $z = z_1$, $z = z_2$, satisfying the equation

(20·28) $$\mu = \mu' \sin \theta'.$$

The ray is a periodic plane curve oscillating between these planes and touching them, the increments in x and y between successive contacts being respectively

(20·29) $$a \int_{z_1}^{z_2} \frac{dz}{(\mu^2 - a^2 - b^2)^{\frac{1}{2}}}, \quad b \int_{z_1}^{z_2} \frac{dz}{(\mu^2 - a^2 - b^2)^{\frac{1}{2}}}.$$

The layer containing the ray may, of course, extend to infinity above or below, in which case the modifications in the argument are obvious.

Let us now consider the case where the refractive index has *spherical symmetry* with respect to a point O. This corresponds approximately to the case of refraction in the earth's atmosphere, when the curvature of the earth is taken into account. Let $\mathbf{r}$ denote the position vector of a point in the medium, relative to O, so that $\mu = \mu(r)$. Then

(20·30) $$\mathbf{i} = \frac{d\mathbf{r}}{ds}, \quad \operatorname{grad} \mu = \mathbf{r} \frac{d\mu}{r \, dr}.$$

Now, by (20·17), we have along a ray

(20·31) $$\frac{d}{ds}(\mu \mathbf{r} \times \mathbf{i}) = \frac{d\mathbf{r}}{ds} \times \mu \mathbf{i} + \mathbf{r} \times \frac{d}{ds}(\mu \mathbf{i})$$
$$= \mathbf{r} \times \operatorname{grad} \mu$$
$$= 0.$$

Thus $\mu \mathbf{r} \times \mathbf{i}$ is a constant vector, which shows that *the ray lies in a plane through O, and further $\mu r \sin \phi = $ const., where ϕ is the angle between the radius vector and the ray*. The analogy to the dynamical theory of orbits under central forces is obvious. This relation may also be written $\mu p = $ const., where p is the perpendicular dropped from O on the tangent to the ray.

Returning to the general heterogeneous medium, let us denote by α, β, γ the direction cosines of a ray, and let us define the *components* σ, τ, υ of a ray by the equations

(20·32) $$\sigma = \mu \alpha, \quad \tau = \mu \beta, \quad \upsilon = \mu \gamma.$$

Then the equations (20·13) for a ray may be written

$$(20·33) \qquad \frac{d\sigma}{ds} = \frac{\partial \mu}{\partial x}, \quad \frac{d\tau}{ds} = \frac{\partial \mu}{\partial y}, \quad \frac{dv}{ds} = \frac{\partial \mu}{\partial z}.$$

We may note that in the case where μ is a function of z only, the result (20·20) may be written $\sigma = $ const., $\tau = $ const.

The law of refraction when a ray passes from one medium to another across a surface of discontinuity of μ may be deduced easily from (20·9). (We might also deduce it from (20·33), by proceeding to a limit in which the gradient of the refractive index

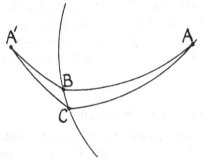

Fig. 37

tends to infinity.) Let $A'BA$ (Fig. 37) be a natural ray, crossing a surface of discontinuity at B, and let $A'CA$ be an adjacent broken curve. In (20·9), which is a general formula for variation of optical length in a single medium, the integral vanishes if the curve from which the variation is made is a natural ray: the formula then reads, if s be taken for parameter,

$$(20·34) \qquad \delta L = \left[\Sigma \mu \alpha \, \delta x \right]_{u_1}^{u_2} = \left[\Sigma \sigma \, \delta x \right]_{u_1}^{u_2}.$$

Applying this formula first to the variation from $A'B$ to $A'C$, and secondly to the variation from BA to CA, we have for differences in optical lengths

$$(20·35) \qquad [A'C] - [A'B] = \Sigma \sigma_2 \delta x,$$
$$[CA] - [BA] = -\Sigma \sigma_1 \delta x,$$

where σ_1, τ_1, v_1 are the components of the ray just *after* refraction, σ_2, τ_2, v_2 the components just *before*, and δx, δy, δz the components of the displacement BC. Consequently (20·34) gives

$$(20·36) \qquad [A'CA] - [A'BA] = -\Sigma\Delta\sigma\,\delta x,$$

where $\Delta\sigma$, $\Delta\tau$, Δv are the increments on crossing the surface. Hence, by Fermat's principle,

$$(20·37) \qquad \Sigma\Delta\sigma\,\delta x = 0$$

for an arbitrary infinitesimal displacement on the surface, and so

$$(20·38) \qquad \Delta\sigma/l = \Delta\tau/m = \Delta v/n,$$

where l, m, n are the direction cosines of the normal to the surface, as in (2·17). The same formulae hold for reflection. *Thus (20·38) is the law of reflection or of refraction at a surface of discontinuity of the refractive index between two heterogeneous isotropic media.*

21. The characteristic function V.

Let us now introduce the *characteristic function V* for a heterogeneous isotropic medium. Let $A'(x', y', z')$ and $A(x, y, z)$ be two arbitrarily selected points in the medium: *the characteristic function*

$$(21·1) \qquad V = V(x', y', z', x, y, z)$$

is defined to be the optical length of the natural ray $A'A$. Let us seek an expression for the infinitesimal change in V due to arbitrary infinitesimal displacements of A' and A to B' and B respectively. If u is a parameter running between the same terminal values on the varied and unvaried rays, the variation in the optical length, that is, δV, is given by (20·9). Since the unvaried curve is a natural ray, the integral vanishes, and if we take $u = s$, the arc-length of the ray, we have

$$(21·2) \qquad \frac{\partial w}{\partial \dot{x}} = \mu\alpha = \sigma, \quad \frac{\partial w}{\partial \dot{y}} = \mu\beta = \tau, \quad \frac{\partial w}{\partial \dot{z}} = \mu\gamma = v.$$

Hence, using accents to denote initial values, we have

$$(21·3) \qquad \delta V = \Sigma\sigma\,\delta x - \Sigma\sigma'\,\delta x',$$

and so†

$$(21\cdot4) \quad \begin{cases} \dfrac{\partial V}{\partial x'} = -\sigma', & \dfrac{\partial V}{\partial y'} = -\tau', & \dfrac{\partial V}{\partial z'} = -v', \\[2ex] \dfrac{\partial V}{\partial x} = \sigma, & \dfrac{\partial V}{\partial y} = \tau, & \dfrac{\partial V}{\partial z} = v. \end{cases}$$

Hence on account of the identities

$$(21\cdot5) \qquad \sigma'^2 + \tau'^2 + v'^2 = \mu'^2, \quad \sigma^2 + \tau^2 + v^2 = \mu^2,$$

we have

$$(21\cdot6) \quad \begin{cases} \left(\dfrac{\partial V}{\partial x'}\right)^2 + \left(\dfrac{\partial V}{\partial y'}\right)^2 + \left(\dfrac{\partial V}{\partial z'}\right)^2 = \mu'^2, \\[2ex] \left(\dfrac{\partial V}{\partial x}\right)^2 + \left(\dfrac{\partial V}{\partial y}\right)^2 + \left(\dfrac{\partial V}{\partial z}\right)^2 = \mu^2. \end{cases}$$

These two partial differential equations‡ are satisfied by the characteristic function V.

If we have a system of media, we define the characteristic function V for the system as the optical length of the natural ray joining a point $A'(x', y', z')$ in the initial medium to a point $A(x, y, z)$ in the final medium. Let us vary A', A to B', B respectively. By Fermat's principle, the optical length of the natural ray $B'B$ is equal (to the first order) to that of a curve C joining B' and B and coinciding with the ray $A'A$ except in the initial and final media. Application of $(20\cdot9)$ to the terminal portions leads us at once to the expression $(21\cdot3)$ for δV, and hence to the equations $(21\cdot4)$ and $(21\cdot6)$, which consequently are true not only for a single medium, but also for a system of media separated by surfaces across which the refractive index is discontinuous.

22. The construction of Huyghens.

Let us now consider the *construction of Huyghens*. We imagine a wave-front S at time t (Fig. 38). To find the wave-front at time $t + dt$, we take elementary spheres having their centres on S with

† It is assumed here that arbitrary independent variations may be given to A' and A: cf. the footnote in connection with $(5\cdot9)$.

‡ Cf. $(5\cdot14)$. Either of the equations $(21\cdot6)$ may be regarded as the Hamilton-Jacobi equation for a particle moving in a conservative field of force.

radii equal to the distance travelled by light in time dt, namely $v\,dt$. The new wave-front is the envelope of these spheres.

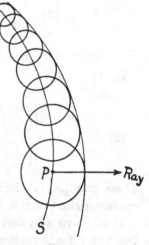

Actually, the envelope consists of two sheets, one in front and one behind S, but we suppose the sense of propagation assigned, so that one of these sheets is ruled out.

The wave of which we are speaking reaches a given point x, y, z at some time t. This t is a function of x, y, z, and so we can write the equation of the wave-front in the form

(22·1) $ct = S(x,y,z).$

This equation describes the whole history of the wave-front. The instantaneous position is given by taking t constant.

Fig. 38

It is evident that, given $v(x,y,z)$, or equivalently $\mu(x,y,z)$, the construction of Huyghens gives a definite development for the wave. In this construction the rays are defined by the condition that the ray through a point P of S passes through the point of contact of the adjacent envelope with the elementary sphere having its centre at P. It is obvious, then, that the ray is normal to the wave. (This is not necessarily the case for anisotropic media, the elementary waves not being spheres.) Accordingly if α, β, γ are the direction cosines of a ray, we have

(22·2) $\theta\alpha = \dfrac{\partial S}{\partial x}, \quad \theta\beta = \dfrac{\partial S}{\partial y}, \quad \theta\gamma = \dfrac{\partial S}{\partial z},$

where θ is a factor of proportionality. By (22·1) we have, moving with the wave,

(22·3) $c\,dt = dS = \Sigma\dfrac{\partial S}{\partial x}dx = \left(\Sigma\alpha\dfrac{\partial S}{\partial x}\right)ds = \theta\,ds,$

where ds is an element of the ray. But $ds = v\,dt$, $\mu = c/v$, and so $\mu = \theta$. Thus by (22·2) we have for the components of the ray, defined as in (20·32),

(22·4) $$\sigma = \frac{\partial S}{\partial x}, \quad \tau = \frac{\partial S}{\partial y}, \quad v = \frac{\partial S}{\partial z}.$$

Therefore S *satisfies the partial differential equation*

(22·5) $$\left(\frac{\partial S}{\partial x}\right)^2 + \left(\frac{\partial S}{\partial y}\right)^2 + \left(\frac{\partial S}{\partial z}\right)^2 = \mu^2.$$

We shall now show that the rays, defined as above in terms of the construction of Huyghens, are in fact identical with the rays given by the principle of Fermat, each being determined by an initial point and direction. Along the ray as given by the construction of Huyghens we have by (22·4)

(22·6) $$\frac{d\sigma}{ds} = \frac{d}{ds}\frac{\partial S}{\partial x}$$

$$= \frac{\partial^2 S}{\partial x^2}\alpha + \frac{\partial^2 S}{\partial x\,\partial y}\beta + \frac{\partial^2 S}{\partial x\,\partial z}\gamma$$

$$= \mu^{-1}\left[\frac{\partial^2 S}{\partial x^2}\frac{\partial S}{\partial x} + \frac{\partial^2 S}{\partial y\,\partial x}\frac{\partial S}{\partial y} + \frac{\partial^2 S}{\partial z\,\partial x}\frac{\partial S}{\partial z}\right]$$

$$= \tfrac{1}{2}\mu^{-1}\frac{\partial}{\partial x}\left[\Sigma\left(\frac{\partial S}{\partial x}\right)^2\right]$$

$$= \tfrac{1}{2}\mu^{-1}\frac{\partial}{\partial x}\mu^2$$

$$= \frac{\partial \mu}{\partial x}.$$

But this is the first of the differential equations (20·33) satisfied by the rays given by the principle of Fermat, and the other two differential equations follow of course similarly. *Thus we are able to reconcile completely the principle of Fermat and the construction of Huyghens in a heterogeneous isotropic medium.*

When we have to deal with reflection or refraction, it is easy to see that the construction of Huyghens gives the same law as Fermat's principle, namely (20·38).

We saw in § 3 that, in the case of homogeneous media, rays emanating from a point source form a normal rectilinear congruence after any number of reflections or refractions. In the case of heterogeneous media, it follows at once from the construction of Huyghens that *rays emanating from a point source form a normal congruence of curves after any number of reflections or refractions*. This result also follows from the formula (21·3) for the variation of the characteristic function, which, as we have seen, holds not only for a single medium, but also for a system of media. For if $A'(x', y', z')$ is a point source in the initial medium and $A(x, y, z)$, $B(x + \delta x, y + \delta y, z + \delta z)$ adjacent points in the final medium such that the optical lengths of the rays $A'A$, $A'B$ have a common value V, then, since

$$\delta x' = \delta y' = \delta z' = 0,$$

and $\delta V = 0$, we have

(22·7) $$\Sigma \sigma \, \delta x = 0,$$

which establishes the orthogonality of the ray $A'A$ to the surface $V = \text{const}$. *It is evident that the surfaces $V = \text{const}$. are in fact the successive positions of a wave.*

The function S which occurs in the interpretation of the construction of Huyghens is closely related to the characteristic function V. As we pass along a ray we have by (22·4)

(22·8) $$dS = \Sigma \frac{\partial S}{\partial x} dx = \Sigma \frac{\partial S}{\partial x} \alpha \, ds = \mu \, ds.$$

Thus the increment in S on passing from one wave to another is the optical length of a ray, measured from one wave to the other. Let S_0 be any wave. Given any point x, y, z, let the ray through it be drawn, cutting S_0 at x', y', z'. Then

(22·9) $$S(x, y, z) - S(x', y', z') = V(x', y', z', x, y, z).$$